Assessment of Department of Defense Basic Research

Committee on Department of Defense Basic Research

Division on Engineering and Physical Sciences

NATIONAL RESEARCH COUNCIL
OF THE NATIONAL ACADEMIES

THE NATIONAL ACADEMIES PRESS
Washington, D.C.
www.nap.edu

THE NATIONAL ACADEMIES PRESS   500 Fifth Street, N.W.   Washington, DC 20001

NOTICE: The project that is the subject of this report was approved by the Governing Board of the National Research Council, whose members are drawn from the councils of the National Academy of Sciences, the National Academy of Engineering, and the Institute of Medicine. The members of the committee responsible for the report were chosen for their special competences and with regard for appropriate balance.

This is a report of work supported by Grant No. N00014-00-G-0230 between the U.S. Navy and the National Academy of Sciences. Any opinions, findings, conclusions, or recommendations expressed in this publication are those of the author(s) and do not necessarily reflect the views of the organizations or agencies that provided support for this project.

International Standard Book Number 0-309-09443-7 (Book)
International Standard Book Number 0-309-54649-4 (PDF)

Limited copies of this report are available from:

Air Force Science and Technology Board
National Research Council
500 Fifth Street, N.W.
Washington, DC 20001
(202) 334-3118

Additional copies are available from:

The National Academies Press
500 Fifth Street, N.W.
Lockbox 285
Washington, DC 20055
(800) 624-6242 or (202) 334-3313
(in the Washington metropolitan area)
Internet, http://www.nap.edu

Copyright 2005 by the National Academy of Sciences. All rights reserved.

Printed in the United States of America

THE NATIONAL ACADEMIES
Advisers to the Nation on Science, Engineering, and Medicine

The **National Academy of Sciences** is a private, nonprofit, self-perpetuating society of distinguished scholars engaged in scientific and engineering research, dedicated to the furtherance of science and technology and to their use for the general welfare. Upon the authority of the charter granted to it by the Congress in 1863, the Academy has a mandate that requires it to advise the federal government on scientific and technical matters. Dr. Bruce M. Alberts is president of the National Academy of Sciences.

The **National Academy of Engineering** was established in 1964, under the charter of the National Academy of Sciences, as a parallel organization of outstanding engineers. It is autonomous in its administration and in the selection of its members, sharing with the National Academy of Sciences the responsibility for advising the federal government. The National Academy of Engineering also sponsors engineering programs aimed at meeting national needs, encourages education and research, and recognizes the superior achievements of engineers. Dr. Wm. A. Wulf is president of the National Academy of Engineering.

The **Institute of Medicine** was established in 1970 by the National Academy of Sciences to secure the services of eminent members of appropriate professions in the examination of policy matters pertaining to the health of the public. The Institute acts under the responsibility given to the National Academy of Sciences by its congressional charter to be an adviser to the federal government and, upon its own initiative, to identify issues of medical care, research, and education. Dr. Harvey V. Fineberg is president of the Institute of Medicine.

The **National Research Council** was organized by the National Academy of Sciences in 1916 to associate the broad community of science and technology with the Academy's purposes of furthering knowledge and advising the federal government. Functioning in accordance with general policies determined by the Academy, the Council has become the principal operating agency of both the National Academy of Sciences and the National Academy of Engineering in providing services to the government, the public, and the scientific and engineering communities. The Council is administered jointly by both Academies and the Institute of Medicine. Dr. Bruce M. Alberts and Dr. Wm. A. Wulf are chair and vice chair, respectively, of the National Research Council.

www.national-academies.org

COMMITTEE ON DEPARTMENT OF DEFENSE BASIC RESEARCH

LARRY D. WELCH, *Chair*, U.S. Air Force (retired), Institute for Defense Analyses, Alexandria, Virginia
C.D. (DAN) MOTE, JR., *Vice Chair*, University of Maryland, College Park
ALBERT J. BACIOCCO, JR., U.S. Navy (retired), Mt. Pleasant, South Carolina
JACK R. BORSTING, University of Southern California, Los Angeles
JOHN M. DEUTCH, Massachusetts Institute of Technology, Cambridge
CHARLES B. DUKE, Xerox Innovation Group, Webster, New York
JOHN S. FOSTER, JR., Northrop Grumman Space Technology (retired), Redondo Beach, California
MARY L. GOOD, University of Arkansas, Little Rock
ROBERT J. HERMANN, Global Technology Partners, Hartford, Connecticut
JAMES C. McGRODDY, IBM Corporation (retired), Briarcliff Manor, New York
C. BRADLEY MOORE, Northwestern University, Evanston, Illinois
JAMES G. O'CONNOR, Pratt & Whitney (retired), Coventry, Connecticut
RICHARD C. POWELL, University of Arizona, Tucson
FAWWAZ T. ULABY, University of Michigan, Ann Arbor
BARBARA A. WILSON, Jet Propulsion Laboratory, Pasadena, California
JOHNNIE E. WILSON, U.S. Army (retired); Dimensions International, Inc., Alexandria, Virginia

Staff

MICHAEL A. CLARKE, Lead Division Board Director
JAMES C. GARCIA, Study Director
LANITA R. JONES, Program Assistant
DANIEL E.J. TALMAGE, JR., Research Associate
LINDSAY D. MILLARD, Intern

Preface

The U.S. Congress mandated that this study be conducted. The mandate is contained in the U.S. Senate report that accompanied the Senate's version of the National Defense Authorization Act for Fiscal Year 2004 and in the U.S. House of Representatives report that accompanied the House-Senate authorization conference committee version of the act. Specifically, the Senate and House reports, respectively, state:

> While the Department [of Defense] is increasing its budget request for the Science and Technology Program, the committee remains concerned that the investment in basic research has remained stagnant and is too focused on near-term demands. Therefore, the committee recommends an increase of $50.0 million for basic research. In addition, the committee directs the Director of Defense Research and Engineering to commission a study by the National Academy of Sciences to assess the basic research portfolio of the services and the Defense Advanced Research Projects Agency (DARPA). This assessment should review the basic research portfolio in order to determine if the programs are consistent with the definitions of basic research in DoD regulation. This report is not intended to rate the worthiness of the basic research portfolio, but rather to determine whether the basic research portfolio needs to be realigned to be more consistent with the goals of traditional fundamental research activities.[1]

---

[1] Senate Armed Services Committee, *FY04 National Defense Authorization Act*, 108th Cong., 2003, S. Rep. 108-46, Title II, Subtitle A. Available online at http://frwebgate.access.gpo.gov/cgi-bin/getdoc.cgi?dbname=108_cong_reports&docid=f:sr046.108.pdf. Last accessed on November 16, 2004.

and

The conferees further note their concerns about funding levels and technical content of the basic research activities of the defense science and technology program. The Department's investment in basic research provides the foundation upon which our modern military is built. It is critical the basic research investment remain strong, stable, and focused on the fundamental search for new knowledge. Therefore, the conferees direct the National Academies of Science to evaluate the DOD basic research portfolio. The evaluation shall utilize the official DOD definition of basic research to determine whether the basic research portfolio is consistent with the definition provided in DOD regulation. The conferees expect to work closely with the National Academies of Science and the Secretary to build the terms of reference for this evaluation. The evaluation should be made available to the congressional defense committees prior to the fiscal year 2006 budget request.[2]

The Department of Defense (DOD) awarded National Research Council (NRC) the study grant with an effective starting date of March 2004.

BACKGROUND AND SCOPE OF STUDY

The Department of Defense currently defines basic research as follows:[3]

Basic research is systematic study directed toward greater knowledge or understanding of the fundamental aspects of phenomena and of observable facts without specific applications towards processes or products in mind. It includes all scientific study and experimentation directed toward increasing fundamental knowledge and understanding in those fields of the physical, engineering, environmental, and life sciences related to longterm national security needs. It is farsighted high payoff research that provides the basis for technological progress. Basic research may lead to: (a) subsequent applied research and advanced technology developments in Defense-related technologies, and (b) new and improved military functional capabilities in areas such as communications, detection, tracking, surveillance, propulsion, mobility, guidance and control, navigation, energy conversion, materials and structures, and personnel support. Program elements in this category involve pre-Milestone A efforts.

---

[2]House Armed Services Committee, *National Defense Authorization Act for Fiscal Year 2004*, 108th Cong., 2003, H.R. Rep. 108-354, Title II, Subtitle D. Available online at http://frwebgate.access.gpo.gov/cgi-bin/getdoc.cgi?dbname=108_cong_reports&docid=f:hr354.108.pdf. Last accessed on November 16, 2004.

[3]Department of Defense, Financial Management Regulation, DOD 7000.14-R, Vol. 2B, Ch. 5, June 2004. Available online at http://www.dtic.mil/descriptivesum/budget_activities.pdf. Last accessed on November 16, 2004.

PREFACE

The goal of DOD basic research support is to encourage advances in fields that are likely to contribute to national defense, and in doing so, to foster a competitive technology base for the U.S. military.

In order to maintain this competitive technology base, the DOD continues to fund basic research. However, over the past 6 years, it has come to the attention of the congressional committees on armed services that basic research funded by the DOD may be changing. Several organizations, including university research departments and defense laboratories, have described areas of concern. They include the following:

- Some research conducted using funds designated specifically for basic research is not, under the DOD's definition, considered basic research;
- Reporting requirements on DOD grants and contracts have become cumbersome and constraining to basic researchers; and
- Basic research funds are handled differently among the Services, which makes the funds, in some cases, difficult to track and monitor.

These concerns prompted the armed services committees to request that the National Academies perform a study regarding the nature of basic research currently being funded by the Department of Defense. The task includes assessing the DOD's basic research portfolio, including that managed by the Office of the Secretary of Defense, the three military departments, and the Defense Advanced Research Projects Agency (DARPA), to determine if the programs in that portfolio are consistent with the definitions of basic research contained in DOD regulations and consistent with the characteristics associated with fundamental research activities. Specifically, the National Academies' statement of task is shown in Box P-1.

STUDY APPROACH AND CONSTRAINTS

The committee (see Appendix A for biographical sketches of members) approached the study in two basic steps, which corresponded to its first two meetings. The committee devoted its first meeting, on May 5-6, 2004, to understanding the DOD definitions for basic and applied research and the characteristics associated with fundamental research and to gathering data and information that would provide insight into the study issue and background from representatives of the research community. During this meeting, the committee received presentations by personnel from the DOD, the Office of Management and Budget (OMB), the National Science Foundation (NSF), and the Department of Energy (DOE) and from a former member of the Senate Committee on Armed Services staff. Representatives of the Association of American Universities (AAU) and the American Association for the Advancement of Science (AAAS) spoke. The associate provost from the Massachusetts Institute of Technology (MIT) and vice

BOX P-1
Statement of Task

In accordance with Senate Report 108-46, Title II, Subtitle A, and House Report 108-354, Title II, Subtitle D, the National Academies will conduct a study to assess the basic research portfolio of the Department of Defense (DoD), including that managed by the Office of the Secretary of Defense (OSD), the three military departments, and the Defense Advanced Research Projects Agency (DARPA), to determine if the programs in that portfolio are consistent with the definitions of basic research contained in DoD regulations and consistent with the characteristics associated with fundamental research activities. To conduct the study, the National Academies will accomplish the following tasks:

1. Form a study committee that possesses knowledge and expertise in the science and technology areas in which DoD basic research is involved; understanding of the differences and relationships between the DoD science and technology (S&T) program categories of basic research (6.1), applied research (6.2), and advanced technology development (6.3); and understanding of DoD financial management and budget regulations and processes that define basic research and govern the categorization of science and technology programs and related budgets as basic research and understand the historical characteristics associated with fundamental research activities.

2. Review the unclassified and classified DoD basic research portfolio through descriptions and documentation of recent, current, and planned programs; discussions with DoD S&T and basic research policy makers, program managers, and intramural and extramural researchers; on-site examination; testimonies from persons with knowledge relevant to the study issues; and other reference information as applicable.

3. Audit the nature of the research to look at fundamental vs. applied orientation; research program review criteria used by the OSD, military departments, and DARPA; any restrictions being placed upon principal investigators; whether broad agency announcements permit truly innovative approaches to be proposed; and other such indicators.

4. Determine if programs in the DoD basic research portfolio are consistent with the definitions of basic research contained in DoD regulations and consistent with the characteristics associated with fundamental research activities. Identify any instances where programs are not consistent with DoD regulations or are not consistent with the characteristics associated with fundamental research activities.

5. Identify any problems that might arise from the definitions themselves or the regulations, policies, or processes implementing the definitions that have a significant bearing on the study issues.

6. Report findings, conclusions, and recommendations regarding the tasks above.

PREFACE   xi

provosts for research from the University of Southern California (USC) and Howard University made presentations. Speakers from Harvard University and George Mason University discussed how research fits into innovation. The list of guest speakers and titles of their presentations at Meeting 1 is provided in Appendix B.

The committee's second meeting, held on May 26-27, 2004, was devoted to reviewing the DOD's basic research program. It included presentations (see Appendix B) by representatives from the U.S. Army, U.S. Navy, U.S. Air Force, DARPA, the Defense Threat Reduction Agency (DTRA), and the Office of the Secretary of Defense (OSD). Army speakers included representatives of the Assistant Secretary of the Army for Acquisition, Logistics, and Technology (ASAALT); the Army Research Laboratory (ARL) (including the Army Research Office [ARO]); the Army Research Institute (ARI); the Medical Research and Materiel Command (MRMC); the Engineer Research and Development Center (ERDC); and the Research, Development, and Engineering Command (RDECOM). The committee received presentations from Navy representatives of the Office of Naval Research (ONR), the Naval Research Laboratory (NRL), and the Naval warfare centers. Representatives of the Air Force Office of Scientific Research (AFOSR) and the Air Force Research Laboratory (AFRL) made presentations. The director of DARPA's Defense Sciences Office represented DARPA. The DOD chemical and biological defense program was discussed by a DTRA representative. An OSD representative discussed the Department of Defense Experimental Program to Stimulate Competitive Research (DEPSCoR).

The committee also conducted several site visits. Committee members visited DARPA, the three main Service laboratories, and the Navy and Air Force offices responsible for managing their respective Service's basic research program. During each visit, committee members met with key organization leadership personnel in addition to one or more groups of researchers and/or research managers. Discussion topics included the DOD definition of basic research; the perceptions of leadership, researchers, and managers about how well their research fits this definition and about characteristics associated with basic research; trends; concerns; and suggested improvements. Appendix C lists the DOD organizations visited.

Committee members also visited and/or interviewed individuals and groups at the universities shown in Appendix C. Each visit included a meeting with the key person responsible for research at the university (usually a vice president or vice provost for research), as well as one or more groups of DOD-sponsored researchers. In addition to the same topics discussed during the DOD site visits, the discussions at the universities addressed the importance of DOD research funding to the university research enterprise (e.g., faculty development and support, the ability to train graduate students, and the impact on the research agenda of individual researchers and the institution). These same topics were discussed during interviews of university research leaders who were not visited in person.

In selecting the universities that it would invite to participate in its meetings, visits, and/or interviews, the committee attempted to include a representative sample of universities receiving DOD research funding. The universities that received DOD basic and applied research funding in fiscal year 2002 are shown in Appendix E. Although it was impossible for the committee to conduct site visits or interviews with research leaders and others at all of these universities or even a major percentage of them, the committee sought to obtain meaningful information regarding the study issue by selecting a sample that received a significant portion of DOD research funding, included research sponsored by all three military Services and DARPA, and was geographically balanced. In all, the committee's site visits and interviews included discussions with approximately 140 people from 7 DOD research organizations and 14 universities.

Constraints on this study were the normal ones experienced by most such studies—schedule and resources. The primary constraint was the requirement expressed by congressional staff members that the study results be available by the end of 2004.

ACKNOWLEDGMENTS

The committee thanks the many organizations and guest speakers that provided excellent support to the committee. The speakers presented information to the committee that had a direct bearing on the study. From the high quality of the presentations, it was obvious that the speakers and others had spent many hours preparing. From the point of view of the committee, this was time well spent. We hope that the speakers, their organizations, the committee's Department of Defense sponsor, and ultimately the readers of this report will agree.

Finally, the committee thanks the NRC staff members who supported the study. Primary among them were Mike Clarke, Jim Garcia, LaNita Jones, Daniel Talmage, and intern Lindsay Millard.

> Larry D. Welch, *Chair*
> Committee on Department of Defense Basic Research

Acknowledgment of Reviewers

This report has been reviewed in draft form by individuals chosen for their diverse perspectives and technical expertise, in accordance with procedures approved by the National Research Council's (NRC's) Report Review Committee. The purpose of this independent review is to provide candid and critical comments that will assist the institution in making its published report as sound as possible and to ensure that the report meets institutional standards for objectivity, evidence, and responsiveness to the study charge. The review comments and draft manuscript remain confidential to protect the integrity of the deliberative process. We wish to thank the following individuals for their review of this report:

Duane Adams, Carnegie Mellon University,
Rita Colwell, University of Maryland,
Anthony J. DeMaria, Coherent-DEOS,
Gerald P. Dinneen, Honeywell, Inc. (retired),
Fernando L. Fernandez, Stevens Institute of Technology,
Ernest Henley, University of Washington,
Kathryn Logan, Georgia Institute of Technology (retired),
John W. Lyons, U.S. Army Research Laboratory (retired),
John B. Mooney, Jr., U.S. Navy (retired),
Theodore Poehler, Johns Hopkins University,
Charles V. Shank, E.O. Lawrence Berkeley National Laboratory,
James Siedow, Duke University,
Pace Vandevender, Sandia National Laboratories, and
Charles Zukoski, University of Illinois.

Although the reviewers listed above have provided many constructive comments and suggestions, they were not asked to endorse the conclusions or recommendations, nor did they see the final draft of the report before its release. The review of this report was overseen by William G. Agnew (NAE), General Motors Corporation (retired). Appointed by the NRC, he was responsible for making certain that an independent examination of this report was carried out in accordance with institutional procedures and that all review comments were carefully considered. Responsibility for the final content of this report rests entirely with the authoring committee and the institution.

Contents

EXECUTIVE SUMMARY 1

ASSESSMENT OF DEPARTMENT OF DEFENSE (DOD) 7
BASIC RESEARCH
    Introduction, 7
    Definitions and Their Role in Managing Basic Research, 8
        Findings, 11
        Recommendations, 11
    Basic Research in the Wider Cycle of Discovery and Technology
        Exploitation, 12
        Findings, 13
        Recommendation, 13
    Multiple Missions, Motivations, and Management Approaches, 14
        Findings, 17
        Recommendations, 18
    The Demand Versus the Supply, 18
        Findings, 23
        Recommendations, 24

APPENDIXES

A   Biographical Sketches of Committee Members, 27
B   Guest Speaker Presentations to the Committee, 37

C DOD Basic Research Organizations and Universities: Committee Site Visits and/or Interviews, 41
D Definitions of Basic, Applied, and Fundamental Research, 44
E Universities That Received Department of Defense 6.1 and 6.2 Funding in Fiscal Year 2002, 51

Executive Summary

The Department of Defense (DOD) supports basic research to advance fundamental knowledge in fields important to national defense. Over the past 6 years, however, several groups have raised concerns about whether the nature of DOD-funded basic research is changing. The concerns include these: Funds are being spent for research that does not fall under DOD's definition of basic research; reporting requirements have become cumbersome and onerous; and basic research is handled differently by the three services. To explore these concerns, the Congress directed DOD to request a study from the National Research Council (NRC) about the nature of the basic research now being funded by the Department. Specifically, the NRC was to determine if the programs in the DOD basic research portfolio are consistent with the DOD definition of basic research and with the characteristics associated with fundamental research.

SUMMARY OF FINDINGS AND RECOMMENDATIONS

No significant quantities of 6.1 funds (basic research) have been directed toward projects that are typical of research funded under categories 6.2 or 6.3. DOD managers are generally successful in assuring that the basic research funded by DOD fits within its definition. That definition, which precludes having "specific applications . . . in mind" for basic research funded under category 6.1, is not, however, a useful criterion for discriminating between basic and applied research. DOD should modify its definition by acknowledging that basic research "has the potential for broad, rather than specific, application" and "may lead to: . . . the discovery of new knowledge that may later lead to more focused advances."

It is also important to note that the need for discovery from basic research does not end once a specific use is identified, but continues through applied research, development, and operations stages. Basic research is not part of a sequential, linear process from basic research, to applied research, to development, and to application. DOD should view basic research, applied research, and development as continuing activities occurring in parallel, with numerous supporting connections throughout the process.

At the same time, there has been a trend within DOD for reduced attention to unfettered exploration in its basic research program. Near-term DOD needs are producing significant pressure to focus basic research in support of those needs. DOD needs to realign the balance of its basic research effort more in favor of unfettered exploration. Senior DOD management should support long-term exploration and discovery and communicate this understanding to its research managers. Long-term, reliable DOD leadership support for basic research depends on a clear understanding of the research's expected value.

The key to effective management of basic research lies in having experienced, empowered managers. DOD's personnel policies should provide for continuity of research management with managers having an adequate level of authority. DOD should also include within the attributes it assigns to the management of its basic research the discovery of new fundamental knowledge, flexibility to modify goals and approaches, freedom to pursue unexpected paths and high-risk research questions, minimum requirements for detailed reporting, open communications, freedom to publish, unrestricted involvement of students and postdoctoral fellows, no restrictions on nationality of researchers, and stable funding.

The breadth and depth of science and technology (S&T) essential to the DOD mission have expanded greatly in the past decade while simultaneously resources provided for basic research have declined significantly. DOD should adjust its basic research allocation to be more in line with its need to pursue research into expanded areas of S&T and to support more unfettered research and new researchers.

Much greater involvement of university researchers is probably essential to meet the demand for new discovery. Acquiring support from DOD, however, is often difficult for many young university researchers. Furthermore, placing export controls on DOD-sponsored 6.1 research disqualifies it from being considered basic research as defined by NSDD-189 and poses a significant threat to the open character of basic research performed in universities. DOD should recognize NSDD-189, the fundamental research exclusion that provides for the unrestricted character of basic research, in its agreements with universities to perform such research.

SPECIFIC FINDINGS AND RECOMMENDATIONS

The committee's findings and recommendations, which appear in the main body of the report with related discussion, are presented below.

Findings

Finding 1. Department of Defense basic research funds under 6.1 have not been directed in significant amounts to support projects typical of 6.2 or 6.3 funding.

Finding 2. Research managers are well motivated and generally successful in focusing 6.1 funding on the discovery of fundamental knowledge in support of the range of Department of Defense needs.

Finding 3. Having specific applications in mind is not a useful criterion for discriminating between basic and applied research.

Finding 4. The set of attributes and desirable characteristics of basic research widely shared among experienced basic research managers can be beneficial in distinguishing between basic and applied research.

Finding 5. The basic research needs of the Department of Defense are complex and do not end when specific applications are identified.

Finding 6. The need for ongoing discovery from basic research can, and usually does, continue through the applied research, system development, and system operation phases.

Finding 7. Included in the range of values expected from basic research in the Department of Defense are (1) discovery arising from unfettered exploration, (2) focused research in response to identified DOD technology needs, and (3) assessment of technical feasibility.

Finding 8. A recent trend in basic research emphasis within the Department of Defense has led to a reduced effort in unfettered exploration, which historically has been a critical enabler of the most important breakthroughs in military capabilities.

Finding 9. Generated by important near-term Department of Defense needs and by limitations in available resources, there is significant pressure to focus DOD basic research more narrowly in support of more specific needs.

Finding 10. Universities, government laboratories, and industry have overlapping roles in basic research: universities primarily address the creation of broad new knowledge and human competencies, and Department of Defense laboratories and industry are more sharply focused on discovery tied more directly to identified DOD needs.

Finding 11. A clear understanding of the value expected from basic research across its full range provides the most reliable assurance of long-term Department of Defense leadership support for the basic research.

Finding 12. A variety of management approaches in the Department of Defense is appropriate to the widely diverse missions and motivations for basic research.

Finding 13. The key to effective management of basic research lies in having experienced and empowered program managers. Current assignment policies and priorities (such as leaving substantial numbers of program manager positions unfilled) are not always consistent with this need, which might result in negative consequences for the effectiveness of basic research management in the long term.

Finding 14. The breadth and depth of the sciences and technologies essential to the Department of Defense mission have greatly expanded over the past decade.

Finding 15. In real terms the resources provided for Department of Defense basic research have declined substantially over the past decade.

Finding 16. The demand for new discovery argues for significantly increased involvement of university researchers. Yet some younger university researchers in the expanded fields of interest to the Department of Defense are often discouraged by the difficulty in acquiring research support from the department.

Finding 17. Recent pressures to apply restrictions on participation and publication through export controls on Department of Defense-sponsored research funded in 6.1 both disqualify it from being considered basic research as defined by National Security Decision Directive 189 and threaten to change fundamentally the open and public character of basic university research. This finding does not apply to research funded in 6.2.

Recommendations

Recommendation 1. The Department of Defense should change its definition of basic research to the following:

EXECUTIVE SUMMARY

Basic research is systematic study directed toward greater knowledge or understanding of the fundamental aspects of phenomena and has the potential for broad, rather than specific, application. It includes all scientific study and experimentation directed toward increasing fundamental knowledge and understanding in those fields of the physical, engineering, environmental, social, and life sciences related to long-term national security needs. It is farsighted high-payoff research that provides the bases for technological progress. Basic research may lead to (a) subsequent applied research and advance technology developments in Defense-related technologies, (b) new and improved military functional capabilities, or (c) the discovery of new knowledge that may later lead to more focused advances in areas relevant to the Department of Defense.

Recommendation 2. The Department of Defense should include the following attributes in its guidance to basic research managers and direct that these attributes be used to characterize 6.1-funded research: a spirit that seeks first and foremost to discover new fundamental understanding, flexibility to modify goals or approaches in the near term based on discovery, freedom to pursue unexpected paths opened by new insights, high-risk research questions with the potential for high payoff in future developments, minimum requirements for detailed reporting, open communications with other researchers and external peers, freedom to publish in journals and present at meetings without restriction and permission, unrestricted involvement of students and postdoctoral candidates, no restrictions on the nationality of researchers, and stable funding for an agreed timetable to carry out the research.

Recommendation 3. The Department of Defense should abandon its view of basic research as being part of a sequential or linear process of research and development (in this view, the results of basic research are handed off to applied research, the results of applied research are handed off to advanced technology development, and so forth). Instead, the DOD should view basic research, applied research, and the other phases of research and development as continuing activities that occur in parallel, with numerous supporting connections among them.

Recommendation 4. The Department of Defense should set the balance of support within 6.1 basic research more in favor of unfettered exploration than of research related to short-term needs.

Recommendation 5. Senior Department of Defense leadership should clearly communicate to research managers its understanding of the need for long-term exploration and discovery.

Recommendation 6. Personnel policies should provide for the needed continuity of research management in order to ensure a cadre of experienced managers capable of exercising the level of authority needed to effectively direct research resources. Further, in light of the reductions in positions reported to the Committee on Department of Defense Basic Research, the Department of Defense should carefully examine the adequacy of the number of basic research management positions.

Recommendation 7. The Department of Defense should redress the imbalance between its current basic research allocation, which has declined critically over the past decade, and its need to better support the expanded areas of technology, the need for increased unfettered basic research, and the support of new researchers.

Recommendation 8. The Department of Defense should, through its funding and policies for university research, encourage increased participation by younger researchers as principal investigators.

Recommendation 9. To avoid weakening the long and fruitful partnership between universities and Department of Defense agencies, DOD agreements and subagreements with universities for basic research should recognize National Security Decision Directive 189, the fundamental research exclusion providing for the open and unrestricted character of basic research. DOD program managers should also explicitly retain the authority to negotiate export compliance clauses out of basic research grants to universities, on the basis of both the program's specific technologies and its objectives.

Assessment of Department of Defense Basic Research

INTRODUCTION

The central issues addressed by the Committee on Department of Defense Basic Research are these: (1) determining if the content of the basic research portfolios managed by the Department of Defense (DOD) is consistent with the DOD definition of basic research and with the characteristics of basic research, (2) evaluating management challenges arising from the definition of basic research, and (3) identifying constraints on basic research in universities and laboratories arising from the definition and its implementation.

To address these issues, the committee engaged in discussions with leaders and managers across the DOD research enterprise and with others who have special interest in the subject. The discussions included two open plenary sessions. A list of the presentations made at these meetings is provided in Appendix B. In addition, to ensure a well-grounded understanding of DOD basic research management and its effect on researchers, the committee interviewed approximately 140 program managers and researchers located at 7 DOD organizations that manage and/or conduct basic research and 14 universities that are among those receiving the largest aggregate of DOD grants and contracts for research. These DOD organizations and universities are listed in Appendix C. Although the committee did not attempt a statistical analysis of the results of these contacts, consistent themes in the responses of interviewees make the anecdotal evidence credible and useful.

The committee held discussions in plenary sessions and at the DOD research organizations and at universities engaged in DOD-sponsored research and reviewed a large number of documents describing the basic research activities of

DOD organizations and DOD-sponsored research in industry and at universities. Even so, at best, this information covered only a sampling of the DOD 6.1 basic research portfolio. Based on that sampling, the committee found reason to question the appropriateness of the classification as 6.1 basic research for only a small percentage of the work. Even in those cases, the issue usually centers on the implication in the DOD definition of basic research (see the following section) that having specific applications in mind is inconsistent with the purposes of basic research. The committee concluded that discussion of that issue is not productive, just as the distinction itself is not useful. Hence, the committee's conclusion is that, while some trends in basic research are undesirable, as discussed elsewhere in this report, there is no evidence of significant misapplication of basic research funding.

In the course of the committee's work, the following four themes emerged and are addressed in the sections below:

- Definitions and Their Role in Managing Basic Research;
- Basic Research in the Wider Cycle of Discovery and Technology Exploitation;
- Multiple Missions, Motivations, and Management Approaches; and
- The Demand Versus the Supply.

DEFINITIONS AND THEIR ROLE IN MANAGING BASIC RESEARCH

On the basis of its plenary sessions and contacts with program managers and researchers, the committee concludes that those responsible for directing and managing basic research in the DOD are well motivated and generally successful in directing basic research resources for purposes appropriate to the DOD definition of basic research: that is, "systematic study directed toward greater knowledge or understanding of the fundamental aspects of phenomena and of observable facts without specific applications towards process or products in mind."[1] Research managers comply generally with the spirit of this definition, although if it was taken literally and researchers had specific applications in mind, their programs would be disqualified from receiving basic research support. Research managers and the committee agree that such a practice would be inappropriate. Hence, although the military departments and defense agencies have the motivation and processes to ensure that 6.1 funding is spent on basic research that has the potential for fundamental discovery, these bodies would not deny 6.1 support for research simply because it would also fund discovery intended for developing

---

[1]Department of Defense, Financial Management Regulation, DOD 7000.14-R, Vol. 2B, Ch. 5, June 2004. Available online at http://www.dtic.mil/descriptivesum/budget_activities.pdf. Last accessed on November 16, 2004.

technology for military needs. If there were such constraints, there would be much less support for basic research.

Furthermore, if a literal interpretation of the DOD definition was applied, the specifics of what the researcher had in mind could be regarded as a key discriminator in determining whether a program was basic research or applied research. Fortunately, the committee found that research managers apply consistent and reasonable judgment on the level of specificity that is appropriate to the purposes of basic research.

The committee concludes that, although managers are able to apply the current definition of basic research effectively to achieve the purposes of basic research, the phrase "without specific applications towards process or products in mind" is not useful either to furthering the purposes of basic research or to helping ensure that 6.1 funding is properly directed.[2] Various motivations for this distinction may exist. However, the view presented to the committee by several senior managers, and strongly reinforced by members of the committee with extensive experience in senior DOD positions, was that this distinction primarily serves the need for a uniform budget and fiscal accounting classification. The committee concludes that this distinction is not a useful research management tool. Ideally it should be possible to convey the purposes of basic research in such a way as to discriminate basic from applied research on the basis of well-understood and accepted principles.

The committee devoted significant time to creating a reasonably simple, straightforward description of basic research and concluded that the combination of slight change in the current DOD definition and a description of characteristics would best serve the needs of effective management of basic research. Accordingly, the first change that the committee suggests is that the opening statement in the DOD definition (see Appendix D for the current definition) be changed to read as follows:

Basic research is systematic study directed toward greater knowledge or understanding of the fundamental aspects of phenomena and has the potential for broad, rather than specific, application.

It is important to note here that this revised opening statement does not suggest that basic research ends when a specific application or set of specific applications is identified. The committee is aware of many instances in which work on a specific application led to expanded basic research that provided further fundamental discoveries with far broader application than what the researcher

---

[2]This assertion is consistent with the concept of use-inspired basic research proposed by Stokes. Donald E. Stokes, *Pasteur's Quadrant: Basic Science and Technological Innovation*, Washington, D.C.: Brookings Institution Press, 1997, pp. 58-89.

had in mind, even as the potential of specific applications emerged. The current definition, however, precludes basic research when specific applications have been identified.[3]

Regarding characteristics of basic research, the committee found that, while there may be differences in detail, there is a fairly strong consensus on a set of characteristics of basic research that help guide research management. The committee found it useful to assemble a list of the most commonly accepted characteristics. The following is such a list—not a set of criteria. Basic research in universities, for example, should accommodate the following:

- A spirit that seeks first and foremost to discover new fundamental understanding,
- Flexibility to modify goals or approaches in the near term based on discovery,
- Freedom to pursue unexpected paths opened by new insights,
- High-risk research questions with the potential for high payoff in future developments,
- Minimum requirements for detailed reporting,
- Open communications with other researchers and external peers,
- Freedom to publish in journals and present at meetings without restriction and permission,
- Unrestricted involvement of students and postdoctoral candidates,
- No restrictions on the nationality of researchers, and
- Stable funding for an agreed timetable to carry out the research.

Some characteristics that are not consistent with the purposes of basic research include the following:

- Inquiry directed to addressing only specified applications,
- Restricted dissemination of results,
- Specific capabilities as research deliverables,
- Short time horizons for reporting, and
- Contractually restricted direction, method, research staff, and problem statement.

---

[3]The complete proposed definition, included in Recommendation 1 in this section, is the current DOD definition of basic research, slightly revised to address what the committee believes is the most serious problem with the current definition. Rather than propose an entirely new definition that might have its own shortcomings, the committee decided that it would be better to recommend the minimum change necessary to the current definition.

Findings

Finding 1. Department of Defense basic research funds under 6.1 have not been directed in significant amounts to support projects typical of 6.2 or 6.3 funding.

Finding 2. Research managers are well motivated and generally successful in focusing 6.1 funding on the discovery of fundamental knowledge in support of the range of Department of Defense needs.

Finding 3. Having specific applications in mind is not a useful criterion for discriminating between basic and applied research.

Finding 4. The set of attributes and desirable characteristics of basic research widely shared among experienced basic research managers can be beneficial in distinguishing between basic and applied research.

Recommendations

Recommendation 1. The Department of Defense should change its definition of basic research to the following:

> *Basic research is systematic study directed toward greater knowledge or understanding of the fundamental aspects of phenomena and has the potential for broad, rather than specific, application. It includes all scientific study and experimentation directed toward increasing fundamental knowledge and understanding in those fields of the physical, engineering, environmental, social, and life sciences related to long-term national security needs. It is farsighted high-payoff research that provides the bases for technological progress. Basic research may lead to (a) subsequent applied research and advance technology developments in Defense-related technologies, (b) new and improved military functional capabilities, or (c) the discovery of new knowledge that may later lead to more focused advances in areas relevant to the Department of Defense.*

Recommendation 2. The Department of Defense should include the following attributes in its guidance to basic research managers and direct that these attributes be used to characterize 6.1-funded research: a spirit that seeks first and foremost to discover new fundamental understanding, flexibility to modify goals or approaches in the near term based on discovery, freedom to pursue unexpected paths opened by new insights, high-risk research questions with the potential for high payoff in future developments, minimum requirements for detailed reporting, open communications with other researchers and external peers, freedom to publish in journals and present at meetings without restriction and permission, unrestricted involvement of students and postdoctoral candidates, no restrictions on the nationality of researchers, and stable funding for an agreed timetable to carry out the research.

BASIC RESEARCH IN THE WIDER CYCLE OF DISCOVERY AND TECHNOLOGY EXPLOITATION

On the basis of anecdotal evidence received during its briefings and discussions, the committee notes that there would be a significant difference in the basic research program of the Department of Defense if a literal interpretation of the current definition of basic research in the DOD regulations were followed, rather than the actual practices common to successful multiple levels of research in the DOD and elsewhere. A literal interpretation could lead to the perception that the levels of research in 6.1, 6.2, 6.3, and so on, are sequential—that is, it could lead to the erroneous view that basic research (funded as 6.1) provides fundamental knowledge that, when it is to be directed at specific applications, transitions to applied research (funded by 6.2), which, when appropriate, transitions to system development (funded by 6.3, 6.4, and so on). This erroneous sequential vision of research is illustrated in Figure 1.

The linear process and sharp lines of demarcation illustrated in Figure 1 may have correctly described the innovation process in the past and may serve some perceived accounting and other important needs in the present. However, that vision is inconsistent with the best practices in the process of discovery and innovation that support the development of new capabilities to meet national security needs. The sequential-and-separate description projects the understanding that technology pushes system development, whereas in practice, system development often pulls science and technology. A more accurate depiction of effective research activity is shown in Figure 2.

As basic research, applied research, and system development proceed in parallel, continuous communication and interaction take place among the levels of

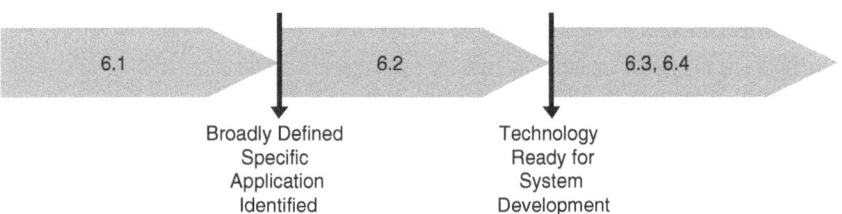

FIGURE 1 Erroneous sequential and separate vision of research.

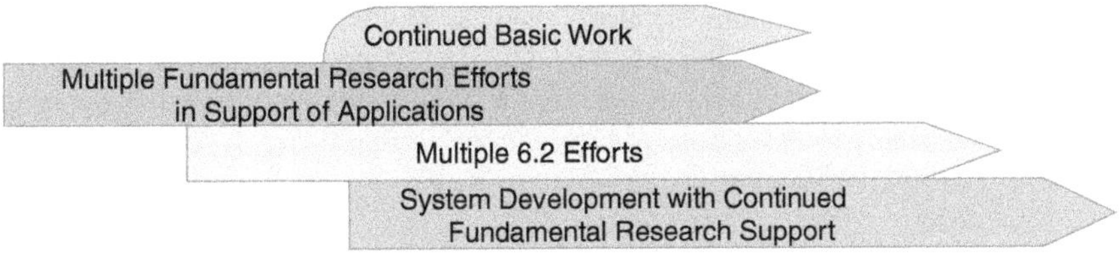

```
            ┌─────────────────────────────┐
            │   Continued Basic Work      ▶
┌───────────────────────────────────────┐
│  Multiple Fundamental Research Efforts │
│       in Support of Applications       ▶
└───────────────────────────────────────┘
      ┌─────────────────────────────────────┐
      │       Multiple 6.2 Efforts          ▶
      └─────────────────────────────────────┘
         ┌──────────────────────────────────┐
         │ System Development with Continued │
         │   Fundamental Research Support    ▶
         └──────────────────────────────────┘
```

FIGURE 2 A more accurate vision: parallel fundamental research.

research. Basic research enables the potential for broadly defined specific applications and continues to contribute until basic and applied research together bring the technology to system development. There is no formal or symbolic handover from basic research to applied research to system development. Even in late stages of system development and testing, issues arise requiring continued or renewed fundamental discovery and applied research. This occurrence is common in the management of industrial product development as well. Product development processes are recognized to involve both reentrant loops to earlier stages and cyclic performance of multiple stages until the product requirements are fully met.[4] Hence, Figure 2 is consistent with current best practices in industrial research and development (R&D).

Findings

Finding 5. The basic research needs of the Department of Defense are complex and do not end when specific applications are identified.

Finding 6. The need for ongoing discovery from basic research can, and usually does, continue through the applied research, system development, and system operation phases.

Recommendation

Recommendation 3. The Department of Defense should abandon its view of basic research as being part of a sequential or linear process of research and development (in this view, the results of basic research are handed off to applied

---

[4]Steven C. Wheelright and Kim B. Clark, *Revolutionizing Product Development: Quantum Leaps in Speed, Efficiency, and Quality*, New York: Free Press, 1992, Ch. 7 (see especially Exhibit 7-4) and Ch. 9 (see especially Exhibit 9-3).

research, the results of applied research are handed off to advanced technology development, and so forth). Instead, the DOD should view basic research, applied research, and the other phases of research as continuing activities that occur in parallel, with numerous supporting connections among them.

MULTIPLE MISSIONS, MOTIVATIONS, AND MANAGEMENT APPROACHES

Department of Defense 6.1-funded research consists of multiple types of activities, and the mix varies over time in response to multiple missions and motivations. The DOD's needs for the fundamental discovery expected of 6.1-funded research are complex and variable, and they will not fit a simple, one-dimensional mold. One type of research need seeks the unfettered[5] exploration of a fundamentally new frontier of knowledge, which, if developed, could have profound influence on military capabilities. Examples of such research include that in nuclear physics in the 1930s and 1940s, in solid-state electronics in the 1950s and 1960s, in photonics in the 1960s to 1980s, and the concept and creation of the ARPAnet as an exploratory, robust communications system in the 1970s. Such high-risk research, although enormously valuable, cannot be the only basic research, even though some part of it leads to new military capabilities that have historically been major sources of U.S. military superiority.

Another type of basic research is the development of standard reference data, such as the properties of materials and their relationship to materials processing. Although this purpose may sound somewhat routine, such information is mandatory for engineering design in the development process and for the assessment of the technical feasibility and cost assessment of incorporating a material in weapons and support systems.

Still another type of basic research constructs exploratory systems or devices that enhance functionality or performance without regard to the design of a robust, cost-effective version. This basic research is usually focused on a well-defined, often near-term technology need that can be relevant to a range of applications.

A consistent and important observation of this committee, based on its interviews, was the current de-emphasis on the first type of basic research described above—the high-risk, high-payoff discovery—and an increased focus on the second and third types. R&D managers find the latter two types of research easier to "justify," given the range of well-defined current needs, whereas the benefits of the first type are more uncertain in the early stages of research. Yet, as noted,

---

[5]The term "unfettered" as used by the committee does not mean unfocused or totally unconstrained. It does mean not being tied to short-term goals or specific applications. It is the kind of research that is truly exploratory and that may or may not, by itself, produce exploitable results. Still, in those cases in which it does not, it is likely to advance knowledge in areas that will have a longer-term payoff.

projects exhibiting the attributes of the first type create the breakthrough benefits for military capabilities. As is the case in private industry,[6] the abnormally large payoffs of a few "big hits" make defense basic research a productive investment overall.

DOD resources committed to basic research create and deliver a portfolio of future valuable returns. It is useful to think of these investments as resulting in options which, if exercised, will lead to solutions for future military challenges as they emerge.[7] The outputs of basic research are not the options themselves, but rather the raw materials used to construct the options in 6.2 and 6.3 research.

Three classes of key research participants generate these raw materials of basic research: universities (receiving about 60 percent of 6.1 funds), government laboratories (receiving about 30 percent), and commercial firms (about 10 percent). Each class has its role in developing the resulting options. Universities create raw materials in the form of new knowledge and human competencies across broad areas of science and engineering. DOD laboratories create new knowledge, but normally in focused areas of importance to DOD applications. Both DOD laboratories and commercial firms have the responsibility of converting the raw materials from basic research into particular technology investment options. The value created by activities conducted by each of these classes can be categorized into several (somewhat overlapping) types that include the following:

- Expansion of the base of technical knowledge underlying the DOD's needs,
- Creation of new technology options,
- Creation of a cadre of technical experts to provide expert advice when needed,
- Recruitment of skilled technical people into the DOD for key positions, and
- Insight into future technology potential and military applications.

These are the values expected from 6.1 investments in basic research. At every level of the R&D chain of command, the values expected should be communicated, so that the sponsors of 6.1 research have a clear and explicit understanding of the value delivered by their investment.

The characteristics, which determine the value of the returns on investment, depend on the broad mission of the investing organization (e.g., the Army, Air

---

[6]Lewis M. Branscomb and Philip E. Auerswald, 2001, *Taking Technical Risks: How Innovators, Executives and Investors Manage High-Tech Risks*, Cambridge, Mass.: MIT Press (see, especially, Ch. 4); and F.M. Scherer and Dietmar Harhoff, 2000, "Technology Policy for a World of Skew-Distributed Outcomes," *Research Policy* 29 (4-5): 559-566. Available online at http://dx.doi.org/doi:10.1016/S0048-7333(99)00089-X. Last accessed on November 16, 2004.

[7]Peter Boer, *The Valuation of Technology: Business and Financial Issues in R&D*, New York, N.Y.: Wiley, 1999; Johnathan Mun, *Real Options Analysis: Tools and Techniques for Valuing Strategic Investments and Decisions*, New York, N.Y.: Wiley, 2002.

Force, Navy, DARPA, and so on). The values expected from a particular investment reflect strongly the mission of the organization performing the research (e.g., a university, government laboratory, or commercial firm). The mix of values created is different for a university than it is for a government laboratory. For a university, the discovery and acquisition of knowledge and the development of skilled personnel are primary values, and they are the values delivered to the DOD for its investment. Research internal to the Service laboratories is focused on exploiting knowledge and human assets to meet military needs. In some cases this work satisfies the "pull" from gaps in the knowledge of science where the DOD has interests that have not attracted the attention of universities and private firms.

To serve the needs of the DOD, the overall values expected from 6.1 investments must be acquired on multiple levels for each Service, laboratory, and investing organization. And the values expected evolve continuously, reflecting the dynamic character of the need for military capabilities. Strategic planning, investment decisions, and retrospective evaluation of the results achieved based on the values expected are the core of an effective management system. Done well, this leads to strong and sustained support for basic research at all levels of the DOD because the investments are well aimed and managed to provide valued returns.

Although the committee believes that the variety of missions, motivations, and management approaches is essential to the range of basic research needs, it is concerned about the clear trend toward increasing short-timescale research in support of near-term applications at the expense of long-term, unfettered exploration of high-risk but potentially large-payoff areas. In one instance among many, the relevance of the proposed work includes potential application to "land vehicle control, sensor networks, control of networks of smart mines and weapon platforms. . . ."[8]

Other indicators of this trend toward sharply focused research are illustrated by DARPA's intent in ensuring a direct connection between 6.1 funding and the specifics of funded projects[9] and by the comment of senior leadership in the Office of Naval Research that "much if not all" of the 6.1 efforts will transition to 6.2 programs.[10] At the same time, the committee found in its site visits and discussions that many research managers and researchers at universities do not know whether 6.1 or 6.2 funds support a particular research effort. This latter fact supports the argument that the ambiguity has not been a serious impediment to attracting university talent and managing their basic research. In any case, given

---

[8]U.S. Army Research Laboratory (ARL), "Funding by Organization During FY 2003," e-mail provided by Carolyn Nash to James Garcia, June 15, 2004.
[9]Committee visit to the Defense Advanced Research Projects Agency, June 15, 2004, Arlington, Va.
[10]Committee visit to the Office of Naval Research, June 21, 2004, Arlington, Va.

the current research limitations discussed in the next section of this report, the reasons for the trend described here are understood but, if it is continued over the long term, it will not serve national security interests. The committee also observed that there are other sources of funding for basic research in which the mission does not drive specific focus as strongly as in the DOD. The National Science Foundation is one of the more notable such sources.

The existence of significant differences in research management approaches within the military departments and defense agencies is consistent with the range of needs. The Air Force, for example, through its Air Force Office of Scientific Research, manages all 6.1 funding in the Air Force, while the Army, Navy, and DARPA manage 6.1 and 6.2 in the same organizations. Furthermore, the Navy manages 6.1 funding centrally in the Office of Naval Research (ONR), while the Army manages 6.1 funding across a number of research organizations. Each of these approaches has strengths and weaknesses, and the committee found no reason to recommend one approach over another. Instead, the committee concludes that the key to effective management of basic research lies in having a cadre of experienced, empowered, and respected 6.1 program managers, supported by uniformly understanding senior leadership deeply committed to basic research. From presentations by DOD research managers, the committee has some concerns relative to the degree of emphasis on maintaining a strong cadre of program managers. In some of the Services, particularly at ONR, substantial numbers of positions have not been refilled when senior people have left. Furthermore, on the basis of the extensive experience of committee members with research in the DOD, the committee feels strongly that an enduring and genuine commitment to basic research needs to be authentic and visible at the Service acquisition executive and senior military levels.

Findings

Finding 7. Included in the range of values expected from basic research in the Department of Defense are (1) discovery arising from unfettered exploration, (2) focused research in response to identified DOD technology needs, and (3) assessment of technical feasibility.

Finding 8. A recent trend in basic research emphasis within the Department of Defense has led to a reduced effort in unfettered exploration, which historically has been a critical enabler of the most important breakthroughs in military capabilities.

Finding 9. Generated by important near-term Department of Defense needs and by limitations in available resources, there is significant pressure to focus DOD basic research more narrowly in support of more specific needs.

Finding 10. Universities, government laboratories, and industry have overlapping roles in basic research: Universities primarily address the creation of broad new knowledge and human competencies, and Department of Defense laboratories and industry are more sharply focused on discovery tied more directly to identified DOD needs.

Finding 11. A clear understanding of the value expected from basic research across its full range provides the most reliable assurance of long-term Department of Defense leadership support for the basic research.

Finding 12. A variety of management approaches in the Department of Defense is appropriate to the widely diverse missions and motivations for basic research.

Finding 13. The key to effective management of basic research lies in having experienced and empowered program managers. Current assignment policies and priorities (such as leaving substantial numbers of program manager positions unfilled) are not always consistent with this need, which might result in negative consequences for the effectiveness of basic research management in the long term.

Recommendations

Recommendation 4. The Department of Defense should set the balance of support within 6.1 basic research more in favor of unfettered exploration than of research related to short-term needs.

Recommendation 5. Senior Department of Defense leadership should clearly communicate to research managers its understanding of the need for long-term exploration and discovery.

Recommendation 6. Personnel policies should provide for the needed continuity of research management in order to ensure a cadre of experienced managers capable of exercising the level of authority needed to effectively direct research resources. Further, in light of the reductions in positions reported to the Committee on Department of Defense Basic Research, the Department of Defense should carefully examine the adequacy of the number of basic research management positions.

THE DEMAND VERSUS THE SUPPLY

The reason for the pressure for more focused basic research at the present time is the intense pressure on all science and technology resources throughout the DOD. Over the past decade, the expectations of military forces have grown to include a far wider range of technologies in far greater depth. The U.S. national military strategy no longer calls for incrementally better capabilities than those of a known adversary. Instead it calls for dominance over a wide range of adversaries in a wide range of circumstances. Innovation is central to underwriting the concepts and supporting the transformation needed to meet these objectives.[11]

The expansion of the range and depth of the DOD's needs should expand the range of researchers and research relevant to those needs. For many reasons, the DOD needs to attract the best and brightest university researchers. The demand for innovation across a wide range of disciplines places a premium on attracting a broad range of research talent. University programs provide proven access to new research vistas, offering new options for meeting challenges. The university programs also give the DOD ready access to eminent scientists and engineers whose talents are needed to address scientific and technical challenges.

To meet these expectations, the expanding range of technologies essential to the DOD mission includes new levels of interest in biological sciences, social sciences, environmental sciences, nanotechnology, robotics, and information technologies. Innovations in these and other areas are as essential to the success of future operations as past innovations in technology were to the success of earlier weapons systems.[12] This circumstance places much greater demand on both basic and applied research. At the same time, pressures on the defense budget are intense with the added costs of transformation and current operations.

Figure 3 shows the change in annual DOD 6.1 funding in real terms (constant dollars) from the 1993 level. The graph shows three lines corresponding to three different sets of inflation indexes (used to convert then-year dollar amounts to base-year, constant-dollar amounts). The figure shows that DOD basic research funding decreased in real terms from 1993 to 1998, then started to increase until 2002, when it began to level out.

As shown in Figure 3, in the face of competing pressures, the 6.1 funding decrease in 2004 from what it was in 1993 was about 10 percent in real terms according to the inflation indexes used by the DOD. The decrease in 2004 was significantly more in real terms if it is calculated using the Consumer Price Index (CPI) or Higher Education Price Index (HEPI) instead of the indexes used by the DOD. Using the CPI, the decrease was about 18 percent. Using the higher education inflation index, it was about 27 percent.

The most common concern expressed by the university community is its perception of shrinking support for university research (corresponding to

---

[11] Office of the Joint Chiefs of Staff, *Joint Vision 2020*, Washington, D.C.: U.S. Government Printing Office, June 2000. Available online at http://www.dtic.mil/jointvision/jvpub2.htm. Last accessed on November 16, 2004.

[12] Defense Science Board, *Defense Science Board Letter Report on DoD Science and Technology Program*, Washington, D.C., August 2000; and DSB, *Report of the Defense Science Board Task Force on Future Strategic Strike Forces*, Washington, D.C., February 2004.

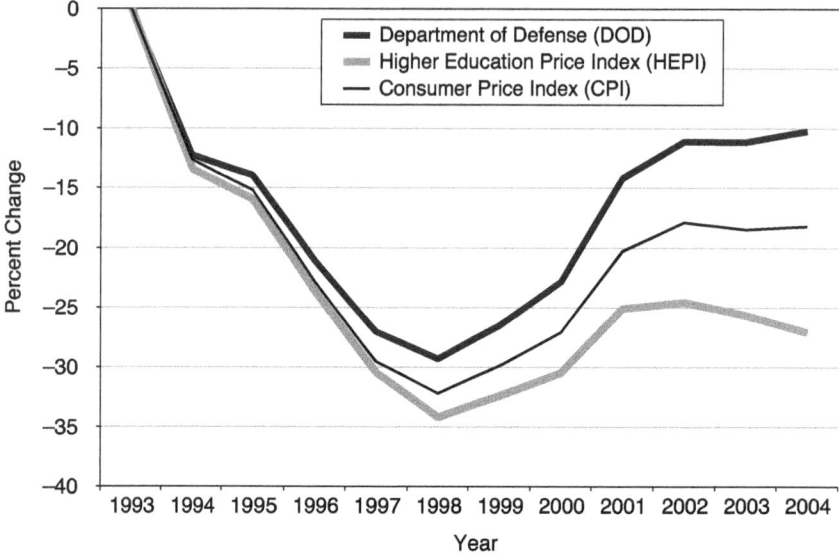

FIGURE 3 Constant-dollar change in annual Department of Defense 6.1 funding (as percentage change from 1993 value).

decreases in DOD 6.1 funding shown in Figure 3). This is broader than the 6.1 funding issue. As noted above, many researchers do not know whether they are funded by 6.1 or by 6.2, and some see an advantage in seeking 6.2 research to attract more substantial funding. University interest in 6.2 should also be welcomed by the DOD and brings many of the same benefits to the DOD as university involvement in 6.1. However, DOD sponsors of 6.2 work are more likely to seek restrictions—such as those on the foreign researcher involvement, the requirement of prepublication review, short time horizons and frequent reporting, and demands for specific findings—that are inconsistent with basic research, and especially with basic research conducted in universities.

University research is also affected by reductions in basic research funding by the states and industry, resulting in more reliance by universities on federal funds. Industry also competes for these funds. A number of the committee's discussions and interviews described the aggregate effects of this environment. While the committee was unable to compile statistically significant data, it found the described effects to be credible. They include the following:

- The funding reduction in 6.1 research is particularly difficult for university engineering and for mathematical and computer sciences. NSF and the DOD are the largest funders of engineering, with the DOD funding about 40 percent, with major concentrations in electrical and mechanical engineering. In these two fields, the DOD provides over half of the federal funding. Regarding computer science and mathematics basic research investment, the DOD funds 17 percent and the Department of Energy funds 3 percent, while NSF funds 75 percent.[13]
- The historical increases in the costs of supporting a graduate student, with no increase in research funding in real terms, results in shorter performance periods and/or fewer graduate students involved in support for DOD needs.
- University research managers indicated that the shrinking support for university research makes it difficult for younger faculty members interested in working on DOD research to get started in DOD research.[14] Consequently, they turn their research attention elsewhere.
- With the increased reporting requirements, principal investigators and graduate students spend more time preparing reports and less time on research.
- The DOD is funding larger grants with industry personnel as the principal investigators.
- Industry funding for basic research has decreased sharply, and many high-tech firms are shifting basic research offshore.

An additional limitation on attracting the best and the brightest to the basic research needs of the United States and the Department of Defense could be the limitation of access to foreign students and scholars. Issues identified during committee discussions and interviews that may have an adverse effect on attracting the best research talent include the following:

- Visa problems that limit the number of foreign students and postdoctoral candidates who are admitted, or who even apply for graduate study in the

---

[13]All percentages are calculated for 2001 based on data found in National Science Foundation, Division of Science Resources Statistics, *Federal Funds for Research and Development, Research to Universities and Colleges by Agency and Field of Science: Fiscal Years 1973-2003*, NSF 04-332, Arlington, Va., 2004. Available online at http://www.nsf.gov/sbe/srs/nsf04332/start.htm. Last accessed on December 2, 2004.

[14]All researchers are affected by funding shortages. During periods of shrinking funding, younger researchers who want to get started in DOD research are at a particular disadvantage relative to those who already have established, ongoing support relationships with DOD research sponsors. With shrinking funds, DOD sponsors' first priority is to continue and complete ongoing research. There is little funding to start new research. Limited funding also means that younger researchers must attract DOD research sponsor support at the expense of support that would have otherwise been given to more established researchers who have already been "proven."

first place (the number of international graduate student applications across the nation has decreased by about one-third from fall 2003 to fall 2004);[15]
- Contracts that prohibit foreign student and postdoctoral candidate participation;
- Recent threats to the exemption of basic research from export controls under National Security Decision Directive (NSDD) 189;
- Possible restrictions placed on the use in research of equipment under export control by foreign students and scholars; and
- Restrictions on foreign student and scholar participation in subcontracts by industry to universities for basic research.

The issue of limitations of access to foreign students and scholars is particularly critical to engineering and the physical, mathematical, and computer sciences. In 1999, 49 percent of all engineering Ph.D. graduates and 47 percent of the mathematics and computer science graduates were foreign nationals (not permanent residents).[16] Those numbers have increased since 1999, and today more than 50 percent of the Ph.D.'s in these areas are on temporary visas. The positive aspect of this problem is that more than 50 percent of these Ph.D.'s remain in the United States.

Recognizing the important contributions of foreign nationals in basic research, President Reagan signed NSDD-189 in 1985, stating that the products of fundamental research should remain unrestricted to the maximum extent possible, and that classification (rather than regulation such as export controls) is the mechanism for control of information.[17] The current Bush administration affirmed that the policy in NSDD-189 "shall remain in effect, and we will ensure that this policy is followed."[18]

---

[15]Heath A. Brown and Peter D. Syverson, *Findings from U.S. Graduate Schools on International Graduate Student Admissions Trends*, Washington, D.C.: Council of Graduate Schools, 2004. Available online at http://www.cgsnet.org/pdf/Sept04FinalIntlAdmissionsSurveyReport.pdf. Last accessed on November 16, 2004.

[16]National Science Board, *Science and Engineering Indicators–2002*, Arlington, Va: National Science Foundation, 2002, NSB-02-1. Available online at http://www.nsf.gov/sbe/srs/seind02/start.htm. Last accessed on November 16, 2004.

[17]National Security Decision Directive 189 was a response to the 1982 report *Scientific Communication and National Security*. The recommendations of the Panel on Scientific Communication and National Security, chaired by Dale Corson, of the Committee on Science, Engineering, and Public Policy, concluded that there is no practical way to restrict international scientific communication without also disrupting domestic scientific information (*Scientific Communication and National Security*, Washington D.C.: National Academy Press, 1982).

[18]Letter from National Security Advisor Condoleeza Rice to Harold Brown, Council on the Future of Technology and Public Policy, November 1, 2001.

However, recent reports from the inspectors general of the DOD[19] and the Department of Commerce[20] appear to restrict foreign student and scholar participation in university basic research. As a result, the joint Association of American Universities (AAU)/Council on Governmental Relations (COGR) Task Force on Restrictions on Research Awards and Troublesome Research Clauses found in April 2004 that "despite affirmations of NSDD-189 by the Administration . . . , troublesome clauses restricting publication and participation by foreign nationals in research awards continue to be a significant problem for universities."[21]

If this trend continues or expands, inadequate access to engineering talent will be a strategic problem. The solution to the problem will, at best, cost the United States dearly; at worst the problem could cost the nation preeminence in vital areas of technical competence.

The net effect of the pressures on resources and the DOD responses to those pressures is that the increase in research resources is not keeping pace with inflation, let alone with the expanded demand for innovation across a broader set of disciplines. It is not surprising, then, that the DOD seeks increased focus in 6.1 research that will support identified capability shortfalls. This emphasis may serve the DOD in the near term, but it certainly will not accommodate the long-term view that is essential to meeting the needs of the department. In short, we are eating our proverbial seed corn.

An additional concern to the committee is the lack of visibility on what happens to 6.1 funding during budget year execution. The committee could find no source of comprehensive information on this subject.

Findings

Finding 14. The breadth and depth of the sciences and technologies essential to the Department of Defense mission have greatly expanded over the past decade.

---

[19]Department of Defense Inspector General, *Report of the Department of Defense Inspector General, Export-Controlled Technology at Contractor, University, and Federally Funded Research and Development Center Facilities,* D-2004-061, Washington, D.C., March 25, 2004. Available online at http://www.dodig.osd.mil/audit/reports/fy04/04-061.pdf. Last accessed on December 2, 2004.

[20]Department of Commerce Inspector General, *Report of the Department of Commerce Inspector General, Deemed Export Controls May Not Stop the Transfer of Sensitive Technology to Foreign Nationals in the U.S.,* IPE-16176, Washington, D.C., March 31, 2004. Available online at http://www.oig.doc.gov/oig/reports/2004/BIS-IPE-16176-03-2004.pdf. Last accessed on December 2, 2004.

[21]Association of American Universities, Council on Government Relations, *Restrictions on Research Awards: Troublesome Clauses: A Report of the AAU/COGR Task Force,* Julie T. Norris, chair. Available online at http://206.151.87.67/docs/Troublesomeclauses.doc. Last accessed on November 16, 2004.

Finding 15. In real terms the resources provided for Department of Defense basic research have declined substantially over the past decade.

Finding 16. The demand for new discovery argues for significantly increased involvement of university researchers. Yet some younger university researchers in the expanded fields of interest to the Department of Defense are often discouraged by the difficulty in acquiring research support from the department.

Finding 17. Recent pressures to apply restrictions on participation and publication through export controls on Department of Defense-sponsored research funded in 6.1 both disqualify it from being considered basic research as defined by National Security Decision Directive 189 and threaten to change fundamentally the open and public character of basic university research. This finding does not apply to research funded in 6.2.

Recommendations

Recommendation 7. The Department of Defense should redress the imbalance between its current basic research allocation, which has declined critically over the past decade, and its need to better support the expanded areas of technology, the need for increased unfettered basic research, and the support of new researchers.

Recommendation 8. The Department of Defense should, through its funding and policies for university research, encourage increased participation by younger researchers as principal investigators.

Recommendation 9. To avoid weakening the long and fruitful partnership between universities and Department of Defense agencies, DOD agreements and subagreements with universities for basic research should recognize National Security Decision Directive 189, the fundamental research exclusion providing for the open and unrestricted character of basic research. DOD program managers should also explicitly retain the authority to negotiate export compliance clauses out of basic research grants to universities, on the basis of both the program's specific technologies and its objectives.

Appendixes

Appendix A

Biographical Sketches of Committee Members

Larry D. Welch (*Chair*) (U.S. Air Force, retired) is senior fellow and immediate past president of the Institute for Defense Analyses (IDA). Prior to joining IDA, he was the 12th chief of staff of the U.S. Air Force, from 1986 to 1990. As chief, he served as the senior uniformed Air Force officer responsible for the organization, training, and equipage of a combined active duty, Guard, Reserve, and civilian force serving at locations in the United States and overseas. As a member of the Joint Chiefs of Staff, he served with the other Service chiefs as the principal military advisers to the Secretary of Defense, National Security Council, and the President. General Welch received a B.A. degree in business administration from the University of Maryland and an M.S. degree in international relations from the George Washington University. He also completed the Armed Forces Staff College and National War College. He was recently awarded the Eugene G. Fubini Award for 2003. This award recognizes highly significant contributions to the Department of Defense in an advisory capacity over a sustained period of time and the providing of expert advice on a diverse range of issues including ballistic missile defense, weapons of mass destruction threats, strategic roadmaps, operational plans, and various transformational technologies.

C.D. (Dan) Mote, Jr. (*Vice Chair*) (NAE) began his tenure as president of the University of Maryland and as Glenn L. Martin Institute Professor of Engineering in September 1998. Prior to assuming the presidency at the University of Maryland, Dr. Mote served on the University of California, Berkeley, faculty for 31 years. From 1991 to 1998, he was vice chancellor at the University of California, Berkeley, and held an endowed chair in mechanical systems. Prior to this, he served as chair of the Department of Mechanical Engineering at Berkeley.

Dr. Mote's research lies in dynamic systems and biomechanics. Internationally recognized for his research on the dynamics of gyroscopic systems and biomechanics, he has authored more than 300 publications; holds patents in the United States, Norway, Finland, and Sweden; and has mentored more than 50 Ph.D. students. He received all his degrees in mechanical engineering from the University of California, Berkeley. Dr. Mote has received numerous awards and honors, including the Humboldt Prize awarded by the Federal Republic of Germany. He is a recipient of the Berkeley Citation and was named Distinguished Engineering Alumnus. He is a member of the American Academy of Arts and Sciences and the National Academy of Engineering and currently serves on its council. He was elected to honorary membership in the American Society of Mechanical Engineers International and is a fellow of the Acoustical Society of America and the American Association for the Advancement of Science. He serves on the Technology Council of Maryland and heads its Technology Transfer Committee of the Greater Washington Board of Trade.

Albert J. Baciocco, Jr. (U.S. Navy, retired) completed his career in the U.S. Navy as a vice admiral in 1987 after 34 years of distinguished service, principally within the nuclear submarine force and directing the Department of the Navy research and technology development enterprise. A graduate of the U.S. Naval Academy in 1953, where he received a Bachelor of Science degree in engineering, he subsequently completed graduate-level studies in nuclear engineering as part of his training for the naval nuclear propulsion program. He served as Chief of Naval Research from 1978 to 1981 and as the director of Research, Development and Acquisition from 1983 to 1987. Since retirement, Admiral Baciocco has been engaged in a broad range of business and pro bono activities with industry, government, and academe, including memberships on the Naval Studies Board and the Army Science Board and on the boards of directors of several corporations, both public and private. He is a trustee of the South Carolina Research Authority and serves as a director of the Research Foundations of the University of South Carolina and the Medical University of South Carolina. He is a member of Tau Beta Pi, a national engineering honor society, and is a recipient of an honorary doctorate in engineering from Florida Atlantic University. Admiral Baciocco has been designated a lifetime national associate of the National Academies by the Council of the National Academies of Sciences.

Jack R. Borsting is professor of business administration and dean emeritus, Marshall School of Business, University of Southern California (USC). From 1994 to September 2001, he served as the executive director of the Center for Telecommunications Management (CTM) at USC, as well as the Morgan Stanley Professor of Business Administration. From 1988 to 1994, Dr. Borsting was dean of USC's School of Business Administration and Robert Dockson Professor of Business Administration. From 1983 to 1988, he was dean of the School of Business Administration at the University of Miami. Previously, Dr. Borsting was Assistant Secretary of Defense (Comptroller) for the U.S. Department of Defense, appointed by Presidents Jimmy Carter and Ronald Reagan. As Comptroller, he acted as chief financial officer for the Secretary of Defense, with overall responsibility for the department's information and budgeting systems, and was a member of the Defense Resources Board. Dr. Borsting has served as provost and academic dean at the Naval Postgraduate School in Monterey, California, and has been the Visiting Distinguished Professor at Oregon State University. He served 2 years with the Air Force as project officer at the Air Force Special Weapons Center in Albuquerque, New Mexico. Dr. Borsting is past president of the Operations Research Society of America and the Military Operations Research Society (MORS), and he is a fellow of the Institute for Operations Research and the Management Sciences, the American Association for the Advancement of Science, the International Engineering Consortium, and MORS. He is past chair and a member of the board of directors of the Los Angeles Orthopaedic Hospital and serves on the advisory council of the Electric Power Research Institute. Dr. Borsting is a trustee of the Rose Hills Foundation and MetLife Investors Trust and also serves on a number of corporate boards. He received his M.A. and Ph.D. degrees in statistics from the University of Oregon and his B.A. degree in mathematics from Oregon State University. He has published articles on operations research and statistics.

John M. Deutch is an Institute Professor at the Massachusetts Institute of Technology (MIT). He served as Director of Central Intelligence from May 1995 to December 1996. From 1994 to 1995, he served as Deputy Secretary of Defense and as Undersecretary of Defense for Acquisition and Technology from 1993 to 1994. Dr. Deutch has also served as Director of Energy Research (1977 to 1979), acting Assistant Secretary for Energy Technology (1979), and Undersecretary (1979 to 1980) in the U.S. Department of Energy. In addition, he has served on the President's Nuclear Safety Oversight Committee (1980 to 1981), the President's Commission on Strategic Forces (1983), the White House Science Council (1985 to 1989), the President's Intelligence Advisory Board (1990 to 1993), the President's Commission on Aviation Safety and Security (1996), and the President's Commission on Reducing and Protecting Government Secrecy (1996). He served as a member of the President's Committee of Advisors on Science and Technology (1997 to 2001) and as chair of the President's Commission to Assess the Organization of the Federal Government to Combat the Proliferation of Weapons of Mass Destruction (1998 to 1999). Dr. Deutch serves as director for the following publicly held companies: Citicorp, Cummins, Raytheon, and Schlumberger Ltd. He has been a member of the MIT faculty since 1970 and has served as chair of the Department of Chemistry, dean of science, and provost. Dr. Deutch has published more than 150 technical publications in physical chemistry, as well as numerous publications on technology, international security, and public policy issues.

Charles B. Duke (NAS/NAE) is vice president and senior research fellow in the Xerox Innovation Group. Prior to holding this position, he was deputy director and chief scientist of the Pacific Northwest Division of the Battelle Memorial Institute and affiliate professor of physics at the University of Washington. From 1972 to 1988 he held various technical and management positions at the Xerox Research Laboratories in Webster, New York, and was an adjunct professor of physics at the University of Rochester. From 1969 to 1972, he was a professor of physics and member of the Materials Research Laboratory and Coordinated Science Laboratory at the University of Illinois in Urbana, following 6 years as a staff member of the General Electric Corporate Research and Development Center in Schenectady, New York. He received his Ph.D. in physics from Princeton University in 1963, following a B.S. degree summa cum laude with distinction in mathematics from Duke University in 1959. He is a fellow and an honorary member of the American Vacuum Society, a fellow of the American Physical Society, a fellow of the Institute of Electrical and Electronics Engineers, a member of the Materials Research Society, and a life member of Sigma Xi. In 1977, Dr. Duke received the Medard W. Welch Award in Vacuum Science and Technology. He served as president of the American Vacuum Society in 1979, on its board of directors for 7 years, and as a trustee from 2003 to 2005. In 1981 he was named one of the ISI 1000 internationally most cited scientists. From 1985 to 1986 he served as founding editor in chief of the *Journal of Materials Research*, and from 1992 to 2001 he was editor in chief of *Surface Science* and *Surface Science Letters*. He was on the council of the Materials Research Society for 7 years, serving as treasurer from 1991 to 1992. In 1993 he was elected to the National Academy of Engineering and in 2001 to the National Academy of Sciences. During the period 1995 to 1999 he served on the council and executive board of the American Physical Society. From 1997 to 2000 he served as general chair of the Physical Electronics Conference. He served on the Governing Board of the American Institute of Physics for 11 years and continues to serve on its Corporate Associates Advisory Committee. He has written more than 350 papers on surface science, materials research, semiconductor physics, and the electronic structure of molecular solids. He holds several patents on the use of feedback in the design of digital imaging and printing systems, wrote a monograph on electron tunneling in solids, and has edited three books: *Surface Science: The First Thirty Years* (1994), *Color Systems Integration* (1998), and *Frontiers in Surface and Interface Science* (2002).

John S. Foster, Jr. (NAE) is chair of the board of GNK Aerospace Transparency Systems; chair of Technology Strategies and Alliances; a member of the board of Wackenhut Services, Inc., and Diana-Hi-Tech; and consultant to Northrop Grumman Space Technology, Sikorsky Aircraft Corporation, Ninesigma, and Defense Group, Inc. He retired from TRW as vice president for science and technology in 1988 and continued to serve on the board of directors of TRW from 1988 to 1994.

Dr. Foster was also Director of Defense Research and Engineering for the Department of Defense, serving for 8 years. He began his career at the Radio Research Laboratory of Harvard University. He spent 2 years as an advisor to the 15th Air Force on radar and radar countermeasures in the Mediterranean theater of operations, and the two summers with the National Research Council of Chalk River, Ontario. Dr. Foster became a division leader in experimental physics at the Lawrence Livermore National Laboratory. He was promoted to associate director, and 3 years later was promoted to director of the Livermore Laboratory and associate director of the Lawrence Berkeley National Laboratory. He received his B.S. degree from McGill University, Montreal, in 1948. He received his Ph.D. in physics from the University of California, Berkeley, in 1952.

Mary L. Good (NAE) is well known for her distinguished career. She has held many high-level positions in academia, industry, and government. The 143,000-member American Association for the Advancement of Science (AAAS) elected Dr. Good to serve as its president following the presidency of Stephen Jay Gould. Dr. Good was the first female winner of the AAAS's Philip Hogue Abelson Prize for outstanding achievements in education, research and development management, and public service, spanning the academic, industrial, and government sectors. Two of her more than 27 awards include the National Science Foundation Distinguished Service medal and the American Chemical Society Priestley Medal. Dr. Good is currently the dean of the Donaghey College of Information Science and Systems Engineering at the University of Arkansas, Little Rock. In addition, she serves as the managing partner of Venture Capital Investors, LLC, in Little Rock. Dr. Good was voted one of Arkansas's Top 100 Women by Arkansas Business. During the terms of Presidents Carter and Reagan, Dr. Good served on the National Science Board and chaired it from 1988 to 1991. She was a member of President George H.W. Bush's Council of Advisors on Science and Technology. Dr. Good was the Undersecretary for Technology in the U.S. Department of Commerce and Technology during President Clinton's first term. This agency assists American industry to advance productivity, technology, and innovation in order to make U.S. companies more competitive in the global market.

Robert J. Hermann (NAE) is currently a senior partner of Global Technology Partners, LLC, a Boston-based investment firm, specializing in investments in technology, defense, aerospace, and related businesses worldwide. In 1998, Dr. Hermann retired from United Technologies Corporation (UTC), where he was senior vice president, science and technology. Prior to joining UTC in 1982, Dr. Hermann served 20 years with the National Security Agency, with assignments in research and development, operations, and NATO. In 1977, he was appointed Principal Deputy Assistant Secretary of Defense for Communications, Command, Control, and Intelligence. In 1979, he was named Assistant Secretary of the Air Force for research, development, and logistics, and in parallel he was director of the National Reconnaissance Office. He received B.S., M.S., and Ph.D. degrees in electrical engineering from Iowa State University. He is currently a member of the following organizations: the Defense Science Board, the National Academy of Engineering, and the board of directors of Orbital Sciences Corporation. His prior organizational memberships include the National Society of Professional Engineers Industry Advisory Group; chair of the Visiting Committee on Advanced Technology of the National Institute of Standards and Technology; board of trustees for the Hartford Graduate Center; chair, co-chair, National Research Council Commission on Physical Sciences, Mathematics, and Applications; the President's Foreign Intelligence Advisory Board; the Commission on the Roles and Missions of the U.S. Intelligence Activities; chair, board of directors of the American National Standards Institute; chair, board of directors of Draper Laboratory; and board of directors, Condor Systems, Inc.

James C. McGroddy (NAE) retired from IBM Corporation as a senior vice president for research at the end of 1996, after leading its research laboratories from 1989 to 1995. During his tenure, which spanned the period of IBM's most difficult challenges, he led a major restructuring of its research efforts, building a model and management system that is now widely emulated. One of the measures of success was the creation during this period of two new laboratories, one in Beijing and one in Austin, Texas. His leadership was recognized by being awarded the Frederik Philips Medal of the IEEE and the George Pake Award of the American Physical Society. He is currently an advisor to several government agencies, a participant in a number of National Research Council groups, and an advisor and a visitor at several universities in the United States and Europe. Dr. McGroddy is the chair of the board of MIQS, a company providing clinical information systems and electronic medical record capability aimed at improving the quality and cost-effectiveness of the care of the chronically ill. As chair of the board of the Stellaris Healthcare Network in 2000 and 2001 and as former chair of the board of Phelps Memorial Hospital Center, he has been heavily involved in the restructuring of the local health care delivery system in Westchester County. He is a director of Paxar, Inc., a New York Stock Exchange traded company, and of Advanced Networks and Services, Inc. He is also a trustee of his alma mater, St. Joseph's University in Philadelphia, as well as a member of the advisory boards of a number of start-up firms and university departments. Dr. McGroddy originally joined IBM in its Research Division in 1965 after receiving a Ph.D. in physics from the University of Maryland. He earned his B.S. in physics from St. Joseph's University in Philadelphia in 1958. In his first years at IBM he focused on research in solid-state physics and electronic devices, and as a result of achievements in these areas was named a fellow of both the Institute of Electrical and Electronics Engineers and of the American Physical Society. In the 1970-1971 academic year, he was a visiting professor of physics at the Danish Technical University. Returning to IBM, he served in a number of management positions in research, development, and manufacturing before being named IBM's director of research in 1989. He is a member of the National Academy of Engineering.

C. Bradley Moore (NAE) is a dynamic leader and internationally recognized chemist. He went to Northwestern University in 2003 from the Ohio State University where, as vice president for research and president of the Ohio State University Research Foundation, he spearheaded dramatic increases in research growth. Improvements in the university's research support services and promotion of multidisciplinary programs across the campus were vital hallmarks of Dr. Moore's tenure at Ohio State. While promoting growth and innovation as vice president for research, he was also a Distinguished Professor of Mathematical and Physical Sciences and a professor of chemistry at Ohio State, where he directed an active research program on molecular energy transfer, chemical reaction dynamics, photochemistry, and spectroscopy. A member of the faculty at the University of California, Berkeley, from 1963 to 2000, Dr. Moore also served as chair of the Chemistry Department and dean of the College of Chemistry. In addition, he was a faculty senior scientist at the Lawrence Berkeley National Laboratory from 1974 to 2000, serving as director of its Chemical Sciences Division from 1998 to 2000. Dr. Moore received his undergraduate degree in chemistry from Harvard University in 1960 and his Ph.D. in chemistry from the University of California, Berkeley, in 1963.

James G. O'Connor (NAE) is the former president of Pratt & Whitney. His 34-year career there started in engineering and included key assignments in customer support, program management, manufacturing operations, and general management. He was involved in both military and commercial programs and businesses. His engineering assignments included development and certification of key commercial engines for the Boeing and Douglas aircraft companies. In early 1981, Dr. O'Connor was named vice president of Pratt & Whitney Commercial Engine Business in East Hartford, Connecticut, responsible for all product support. In 1982, he joined Pratt & Whitney's Government Engine Business in West Palm Beach, Florida, as senior vice president for the F100 engine program. He was appointed executive vice president of the Government Engine Business in January 1984 and assumed the post of president in March 1985. In this position he was responsible for all aspects of Pratt's $2 billion business with the U.S. government and 15 foreign governments. In October 1987, Dr. O'Connor returned to East Hartford and was named vice president of manufacturing operations. In 1989 he became the chief executive for Pratt & Whitney. He was responsible for all of the aircraft engine manufacturer's $7 billion operations. He retired in 1993. He is currently chair of the board of trustees of Embry-Riddle Aeronautical University. In addition, he is a member of the National Academy of Engineering, the Connecticut Academy of Science and Engineering, the President's Advisory Council at Clemson University, and the Wings Club.

Richard C. Powell is currently vice president for research and graduate studies and professor at the Optical Sciences Center, University of Arizona. He received his B.S. in physics in 1962 from the U.S. Naval Academy, Annapolis, Maryland; his Ph.D. and M.S. in physics in 1964 and 1967 from Arizona State University, Tempe. From 1964 to 1968, Dr. Powell was a staff scientist at the Air Force Cambridge Research Laboratories in Bedford, Massachusetts, where he worked on the development of new solid-state laser materials and radiation damage in semiconductor devices. Between 1968 and 1971, Dr. Powell was a staff scientist at the Sandia National Laboratories in Albuquerque, New Mexico, where his research involved exciton dynamics in organic crystals and polymers and saturation effects in plastic scintillators. In 1971, Dr. Powell moved to Oklahoma State University in Stillwater, where he was a professor in the Physics Department and director of the Center for Laser Research until 1992. He also served as head of the Physics Department and associate dean of the College of Arts and Sciences. During that period, his research involved laser spectroscopy of solids for use in lasers and nonlinear optics applications. In addition, he participated in several projects involving laser applications in medicine. Dr. Powell also had several temporary assignments, including positions at Motorola Semiconductor Division, the California Institute of Technology, and the Lawrence Livermore National Laboratory. During his scientific career, he has published more than 200 articles, two books, and participated in many national and international conferences.

Fawwaz T. Ulaby (NAE) is the vice president for research and the R. Jamison and Betty Williams Professor of Electrical Engineering and Computer Science at the University of Michigan. He is a member of the National Academy of Engineering and serves on several national scientific boards and commissions. Since joining the University of Michigan faculty in 1984, Professor Ulaby has been directing large, interdisciplinary National Aeronautics and Space Administration (NASA) projects aimed at the development of high-resolution satellite radar sensors for mapping Earth's terrestrial environment. He also served as the founding director of the NASA-funded Center for Space Terahertz Technology. The center's research focuses on the development of microelectronic devices and circuits that operate at wavelengths intermediate between the infrared and the microwave regions of the electromagnetic spectrum. Professor Ulaby has authored eight books, contributed chapters to several others, and published more than 600 scientific papers and reports. His recent undergraduate textbook, *Applied Electromagnetics*, published by Prentice-Hall in January 1997, has been adopted by some 80 universities across the United States. Professor Ulaby is the recipient of numerous awards, including the Eta Kappa Nu Association C. Holmes MacDonald Award as "an Outstanding Electrical Engineering Professor in the United States of America for 1975," the Institute of Electrical and Electronics Engineers (IEEE) Centennial Medal (1984), the American Society of Photogrammetry's Presidential Citation for Meritorious Service (1984), the Kuwait Prize in applied science

(1986), the NASA Group Achievement Award (1990), the University of Michigan's Distinguished Faculty Achievement Award (1991), the University of Michigan Regents Medal for Meritorious Service (1996), the IEEE Millennium Medal for Outstanding Achievements and Contributions (2000), and the 2001 IEEE Electromagnetics Award. Over his 30-year academic career, Professor Ulaby has supervised more than 100 M.S. and Ph.D. graduate students and served as principal investigator on about $40 million in research grants and contracts. In January 2001 he assumed the position of editor in chief of the *IEEE Proceedings*, the most highly cited journal in electrical and computer engineering. In 2002 he received the William Pecora Award, a joint recognition by NASA and the Department of the Interior.

Barbara A. Wilson is currently a program manager in the Solar System Exploration Programs Directorate at the NASA Jet Propulsion Laboratory (JPL), where she manages the development of communications, computing, electronics, and imaging technologies for NASA's Office of Exploration Systems. From 2001 to 2003, she served as the chief technologist of the Air Force Research Laboratory under an Intergovernmental Personnel Act loan agreement between NASA and the Air Force, and as JPL's chief technologist from 1999 to 2001. After earning her Ph.D. in physics from the University of Wisconsin-Madison in 1978, she worked in basic research at AT&T Bell Laboratories, with a focus on quantum structures. Her research contributions were recognized with an AT&T Exceptional Contribution Award. She moved to JPL in 1988, where she has also served as director of the Center for Space Microelectronics, as manager of the Microdevices Laboratory, and as deputy manager and chief technologist of NASA's New Millennium flight validation program. Her leadership in the New Millennium Program earned her both JPL and NASA achievement awards. Dr. Wilson is a fellow of the American Physical Society (APS) and former general councilor and member of the APS Executive Board. She was appointed to the International Academy of Astronautics in 2000. She is currently serving her second term on the Air Force Scientific Advisory Board (SAB). As an SAB member, she has participated in a number of U.S. Air Force studies and was appointed science and technology chair of the SAB in 2004. In this capacity she will lead the external review of the Air Force science and technology programs. Dr. Wilson has also served on numerous other National Science Foundation, National Research Council, and NASA panels.

Johnnie E. Wilson (U.S. Army, retired) is the president and chief operating officer of Dimensions International, Inc. (DI), an information technology company specializing in information integration and providing solutions for the acquisition, analysis, management, and transformation of data into information. His primary responsibility is on the program side, providing oversight to the technical directors and program managers. He assists them in managing, marketing, and expanding their operations. His extensive network, both military and civilian, is a great asset in maturing opportunities that will enhance DI's growth and development. General Wilson entered the Army in August 1961 as an enlisted soldier, attaining the rank of SSG before attending Officer Candidate School (OCS). On completion of OCS in 1967, he was commissioned a second lieutenant in the Ordnance Corps. He was awarded a B.S. degree in business administration from the University of Nebraska at Omaha. General Wilson also holds an M.S. degree in logistics management from the Florida Institute of Technology. Additionally, his military education includes completion of the Ordnance Officer Basic and Advanced Courses, the Army Command and General Staff College, and the Industrial College of the Armed Forces. General Wilson has held a wide variety of important command and staff positions, culminating in his last assignment as the Commanding General, U.S. Army Materiel Command (AMC), an organization of 80,000 people serving throughout the world. As the commanding general of AMC, he was responsible for the Army's wholesale logistics, acquisition, and technology generation operations. As a result, General Wilson possesses extensive knowledge in supply-chain management, acquisition reform, and strategic logistics planning. General Wilson also served as the deputy commanding general, 21st Theater Army Area Command (TAACOM), the Army's largest and most diverse logistics unit. Based on his wide experience with leading soldiers, General Wilson was selected to command the Ordnance Center and School responsible for the training and professional development of thousands of soldiers, noncommissioned officers, and officers every year. Following this successful assignment, he served as the chief of staff, AMC, where he was responsible for resource and personnel management for a workforce with more than 80,000 military and civilian members. From 1994 to 1996, General Wilson served as the Deputy Chief of Staff for Logistics, Department of the Army, where he was responsible for worldwide logistics.

Appendix B

Guest Speaker Presentations to the Committee

The Committee on Department of Defense Basic Research conducted two meetings in May 2004 at which it received presentations by invited speakers. The titles of the presentations and speakers are listed in this appendix. (See Appendix C for a list of committee interviews and site visits to Department of Defense [DOD] basic research organizations and universities.)

The committee devoted its first meeting, on May 5-6, 2004, to understanding the DOD definitions for basic and applied research and the characteristics associated with fundamental research and to gathering data and information relevant to the study from representatives of the research community.

The second meeting, held on May 26-27, 2004, was devoted to reviewing the DOD's basic research program. Presentations were made by representatives of the Army, Navy, Air Force, Defense Advanced Research Projects Agency, Defense Threat Reduction Agency, and the Office of the Secretary of Defense.

MEETING 1, WASHINGTON, D.C., MAY 5-6, 2004

"Department of Defense S&T Management Definitions/Policies Regarding Basic Research"
William O. Berry, Director for Basic Research
Office of the Director of Defense Research and Engineering

"Background Leading to Congressional Tasking"
Carolyn Hanna, Special Assistant for Science and Technology Community Affairs
Department of Homeland Security
Former staff member, Senate Committee on Armed Services

"Department of Defense Basic Research Funding: Some Personal Experiences and Reflections"
Greg Voth, Professor of Chemistry
University of Utah
Chair, Public Policy Committee, American Chemical Society

"Some OMB Thoughts on Proper Counting of Department of Defense Basic Research"
Greg Henry, Senior Program Examiner and Operations Research Analyst
National Security Division, Office Management and Budget

"The Federal Investment in Defense Research"
Kei Koizumi, Director of Research and Development Budget and Policy Program
American Association for the Advancement of Science

"Association of American Universities Insights"
Kenneth F. Galloway, Dean of the School of Engineering and Professor of Electrical Engineering
Vanderbilt University

"DOD Comptroller Definitions/Policies Regarding Basic Research"
Caral Spangler, Director for Investment
Office of the Under Secretary of Defense (Comptroller)

"Basic Research: The Leading Edge of Science and Engineering"
Mary E. Clutter, Assistant Director for Biological Sciences
National Science Foundation

"Basic Research and the Office of Science Insights"
Walt Stevens, Director, Chemical Sciences, Geosciences, and Biosciences Division, Office of Science
Department of Energy

"Department of Defense Sponsored Basic Research"
Claude R. Canizares, Associate Provost, and Bruno Rossi Professor of Experimental Physics
Massachusetts Institute of Technology

"University of Southern California Insight on Basic Research"
Cornelius W. Sullivan, Vice Provost for Research and Professor of Biological Sciences
University of Southern California

"Department of Defense Research and the Historically Black College and Universities Community: Challenges and Opportunities"
Orlando Taylor, Vice Provost for Research, Dean, Graduate School, and Professor, School of Communications
Howard University

"6.1–6.X?"
Robert A. Frosch, Senior Research Associate
Belfer Center for Science and International Affairs
Harvard University

"6.1, 6.2, 6.3? The End of 'R&D'"
Philip Auerswald, Director, Center for Science and Public Policy, and Assistant Professor of Public Policy
George Mason University

"What DOD Wants to Get from Study"
William O. Berry, Director of Basic Research
Office of the Director of Defense Research and Engineering

MEETING 2, WASHINGTON, D.C., MAY 26-27, 2004

"Army Basic Research . . . Accelerating the Pace of Transformation"
John Parmentola, Director for Research and Laboratory Management
Office of the Assistant Secretary of the Army for Acquisition, Logistics, and Technology

"Army Research Laboratory Basic Research"
John Pellegrino, Acting Deputy Director
Army Research Laboratory

"Army Medical Research and Materiel Command Basic Research Overview"
Col. James Romano, Deputy Commander
Army Medical Research and Materiel Command

"Army Research Institute for the Behavioral Sciences Basic Research"
Jonathan Kaplan, Program Manager
Army Research Institute for the Behavioral Sciences

"Army Corps of Engineers: Engineer Research and Development Center Basic Research"
Rick Morrison, Deputy Director
Army Corps of Engineers, Engineer Research and Development Center

"Basic Research at Army Research, Development and Engineering Centers"
Robin Keesee, Acting Deputy Director
Army Research, Development, and Engineering Command

"Air Force Basic Research"
Lyle Schwartz, Director
Air Force Office of Scientific Research

"Basic Research at Air Force Research Laboratory Technical Directorates"
Barry Farmer, Chief Scientist, Materials and Manufacturing Directorate
Air Force Research Laboratory

"Navy S&T Overview and Introduction"
Stephen Lubard, Technical Director, Science and Technology
Office of Naval Research

"Naval Basic Research—Office of Naval Research Contributions"
James Murday, Chief Scientist
Office of Naval Research

"Naval Basic Research—Naval Research Laboratory Contributions"
Bhakta Rath, Associate Director of Research, Materials Science, and Component
    Technology Directorate
Naval Research Laboratory

"Naval S&T—Warfare Center Contributions"
Robert Kavetsky, Warfare Center Liaison to Office of Naval Research
Office of Naval Research

"Defense Experimental Program to Stimulate Competitive Research"
Keith Thompson, Manager, Defense Experimental Program to Stimulate Competitive Research
Office of the Director of Defense Research and Engineering

"Overview of DARPA Basic Research Program"
Steven G. Wax, Director, Defense Sciences Office
Defense Advanced Research Projects Agency

"Chemical and Biological Defense Program Basic Research"
Charles Gallaway, Director, Chemical and Biological Defense
Defense Threat Reduction Agency

Appendix C

DOD Basic Research Organizations and Universities: Committee Site Visits and/or Interviews

The site visits and interviews conducted by the Committee on Department of Defense Basic Research included discussions with approximately 140 people from 7 Department of Defense (DOD) research organizations and 14 universities, as listed in this appendix. (See Appendix B for a list of guest speaker presentations to the committee at its two meetings in Washington, D.C., in May 2004.)

DOD BASIC RESEARCH ORGANIZATIONS VISITED

The committee conducted site visits at the seven (including the Army Research Office) DOD basic research organizations listed below. Committee members met with key organization leadership in addition to one or more groups of researchers and/or research managers at each site. Discussion topics included the DOD definition of basic research; the perceptions of those in leadership positions, researchers, and managers about how well their research fits this definition and about characteristics associated with basic research; trends; concerns; and suggested improvements. The basic research organizations visited are these:

- Air Force Office of Scientific Research (AFOSR)
  Arlington, Virginia

- Air Force Research Laboratory (AFRL)
  Wright-Patterson Air Force Base, Ohio
  (included video conference participation by several AFRL directorates across the United States)

- Army Research Laboratory (ARL)
  Adelphi, Maryland
  (included participants from Army Research Office [ARO], Research Triangle Park, North Carolina)

- Defense Advanced Research Projects Agency (DARPA)
  Arlington, Virginia

- Naval Research Laboratory (NRL)
  Washington, D.C.

- Office of Naval Research (ONR)
  Arlington, Virginia

VISITS AND/OR INTERVIEWS AT UNIVERSITIES

Committee members made site visits and/or conducted interviews with research leaders of 14 universities. Each visit included a meeting with the key person responsible for research at the university, as well as one or more groups of DOD-sponsored researchers. In addition to the same topics addressed during the DOD site visits, these discussions addressed the importance of DOD research funding to the university research enterprise. (These same topics were discussed during interviews of university research leaders who were not visited in person.) The 14 universities are these:

- Cornell University
  Ithaca, New York

- Massachusetts Institute of Technology Institute for Soldier Nanotechnologies
  Cambridge

- North Carolina State University
  Raleigh

- Northwestern University
  Evanston, Illinois

- Ohio State University
  Columbus

- Pennsylvania State University
  State College

APPENDIX C

- University of Arizona
  Tucson

- University of California, Los Angeles

- University of California, Santa Barbara

- University of Michigan
  Ann Arbor

- University of New Mexico
  Albuquerque

- University of Southern California Information Sciences Institute
  Los Angeles

- University of Texas at Austin

- University of Washington
  Seattle

APPENDIX D

Definitions of Basic, Applied, and Fundamental Research

This appendix contains definitions of basic, applied, and fundamental research quoted from various sources.

BASIC RESEARCH

DOD Financial Management Regulation, DOD 7000.14-R, Vol. 2B, Ch. 5:
Basic research is systematic study directed toward greater knowledge or understanding of the fundamental aspects of phenomena and of observable facts without specific applications towards processes or products in mind. It includes all scientific study and experimentation directed toward increasing fundamental knowledge and understanding in those fields of the physical, engineering, environmental, and life sciences related to long-term national security needs. It is farsighted high payoff research that provides the basis for technological progress. Basic research may lead to: (a) subsequent applied research and advanced technology developments in Defense-related technologies, and (b) new and improved military functional capabilities in areas such as communications, detection, tracking, surveillance, propulsion, mobility, guidance and control, navigation, energy conversion, materials and structures, and personnel support. Program elements in this category involve pre-Milestone A efforts. Available online at http://www.dod.mil/comptroller/fmr/02b/Chapter05.pdf. Last accessed on November 16, 2004.

The objective of basic research is to gain more comprehensive knowledge or understanding of the subject under study, without specific applications in mind. In industry, basic research is defined as research that advances scientific knowledge but does not have specific immediate commercial objectives, although it

APPENDIX D

may be in fields of present or potential commercial interest. [National Science Foundation, Directorate for Social, Behavioral & Economic Sciences, US definitions for resource surveys, 1996.] Available online at http://www.nsf.gov/sbe/srs/seind96/ch4_defn.htm. Last accessed on November 16, 2004.

Scientific efforts that seek to gain more comprehensive knowledge or understanding of the subject under study, without specific applications or commercial objectives in mind. Available online at http://energytrends.pnl.gov/glosn_z.htm. Last accessed on November 16, 2004.

Basic research analyzes properties, structures, and relationships toward formulating and testing hypotheses, theories, or laws. As used in this survey, industrial basic research is the pursuit of new scientific knowledge or understanding that does not have specific immediate commercial objectives, although it may be in fields of present or potential commercial interest. Available online at http://caspar.nsf.gov/nsf/srs/IndRD/glossary.htm. Last accessed on November 16, 2004.

The investigation of the natural phenomena as contrasted with applied research. Available online at http://www.onlineethics.org/glossary.html. Last accessed on November 16, 2004.

Systematic study directed toward greater knowledge or understanding of the fundamental aspects of phenomena and of observable facts without specific applications towards processes or products in mind. [OMB Circular A-11, June 1996.] See Conduct of Research and Development. Available online at https://radius.rand.org/radius/demo/glossary.html. Last accessed on November 16, 2004.

Fundamental scientific inquiry to understand the unknown and contribute to improved general knowledge (cf. with applied research). Available online at http://www.ipmrc.com/lib/glossary.shtml. Last accessed on November 16, 2004.

Research done to further knowledge for knowledge's sake. Available online at http://www.modernhumanorigins.com/b.html. Last accessed on November 16, 2004.

Fundamental research; it often produces a wide range of applications, but the output of basic research itself usually is not of direct commercial value. The output is knowledge, rather than a product; it typically cannot be patented. Available online at http://www.wwnorton.com/stiglitzwalsh/economics/glossary.htm. Last accessed on November 16, 2004.

Research aimed at expanding knowledge rather than solving a specific, pragmatic problem. Available online at https://www.quirks.com/resources/glossary.asp. Last accessed on November 16, 2004.

Focused, systematic study and investigation undertaken to discover new knowledge or interpretations and establish facts or principles in a particular field. See Research. Available online at http://www.siu.edu/orda/general/glossary.html. Last accessed on November 16, 2004.

Fundamental research; it often produces a wide range of applications, but the output of basic research itself usually is not of direct commercial value; the output is knowledge, rather than a product; the output of basic research typically cannot be patented. Available online at http://wellspring.isinj.com/sample/econ/micro/glossb.htm. Last accessed on November 16, 2004.

Research emphasizing the solution of theoretical problems. Binomial probability distribution: The probabilities associated with every possible outcome of an experiment involving n independent trials and a success or failure on each trial. Bivariate analysis: The analysis of relationships among pairs of variables. Available online at http://www.prm.nau.edu/prm447/definitions.htm. Last accessed on November 16, 2004.

Basic research is research undertaken to advance the knowledge of methodologies and techniques of research. (Compare applied research.) Available online at http://www.rigneyassoc.com/glossary.html. Last accessed on November 16, 2004.

Research that is directed at the growth of scientific knowledge, without any near-term expectations of commercial applications. Available online at http://highered.mcgraw-hill.com/sites/0072443901/student_view0/chapter4/glossary.html. Last accessed on November 16, 2004.

Research which adds something new to the body of knowledge of a particular field. Available online at http://researchoffice.astate.edu/glossary_of_proposal_terms.htm. Last accessed on November 16, 2004.

Designed to test and refine theory. The purpose is to increase our knowledge about communication phenomena by testing, refining, and elaborating theory. Available online at http://www.uky.edu/~drlane/cohort/define.htm. Last accessed on November 16, 2004.

The purpose is to increase knowledge without concern for practical application. Available online at http://www.ied.edu.hk/csnsie/ar/chap1/1_glossary.htm. Last accessed on November 16, 2004.

In basic research the objective of the sponsoring agency is to gain more complete knowledge or understanding of the fundamental aspects of phenomena and of observable facts, without specific applications toward processes or products in mind. Available online at http://www.nsf.gov/sbe/srs/fedfunds/glossary/def.htm. Last accessed on November 16, 2004.

NSF Definition of Basic Research: Basic research is defined as systematic study directed toward fuller knowledge or understanding of the fundamental aspects of phenomena and of observable facts without specific applications towards processes or products in mind. (In Bill Berry's presentation, Meeting 1, Committee on Department of Defense Basic Research.)

In basic research the objective of the sponsoring agency is to gain fuller knowledge or understanding of the fundamental aspects of phenomena and of observable facts without specific applications toward processes or products in mind. Available online at http://www.aaas.org/spp/cstc/pne/pubs/regrep/alaska/appendices.htm. Last accessed on November 16, 2004.

OMB (Circular A-11, 2003): Basic research is defined as systematic study directed toward fuller knowledge or understanding of the fundamental aspects of phenomena and of observable facts without specific applications towards processes or products in mind.

Research directed toward the increase of knowledge, the primary aim being a greater knowledge or understanding of the subject under study. Available online at http://usmilitary.about.com/library/glossary/b/bldef00823.htm. Last accessed on November 16, 2004.

Commission of the European Communities: While there is no strict, unanimously accepted definition of what constitutes basic research, in practice one can identify and distinguish from other types of research, those which are carried out with no direct link to a given application and, if not exclusively, in any case and above all with the objective of progressing knowledge. Available online at http://europa.eu.int/comm/research/press/2004/pdf/acte_en_version_final_15janv_04.pdf. Last accessed on November 16, 2004.

APPLIED RESEARCH

OMB (Circular A-11, 2003): Applied research is defined as systematic study to gain knowledge or understanding necessary to determine the means by which a recognized and specific need may be met.

The investigation of some phenomena to discover whether its properties are appropriate to a particular need or want. In contrast, basic research investigates phenomena without reference to particular human needs and wants. Available online at http://www.onlineethics.org/glossary.html. Last accessed on November 16, 2004.

Aimed at gaining knowledge or understanding to determine the means by which a specific, recognized need may be met. In industry, applied research includes investigations oriented to discovering new scientific knowledge that has specific commercial objectives with respect to products, processes, or services. [National Science Foundation, Directorate for Social, Behavioral & Economic Sciences, US definitions for resource surveys, 1996.] Available online at http://www.nsf.gov/sbe/srs/seind96/ch4_defn.htm. Last accessed on November 16, 2004.

Applied research is undertaken either to determine possible uses for the findings of basic research or to determine new ways of achieving some specific, predetermined objectives. As used in this survey, industrial applied research is investigation that may use findings of basic research toward discovering new scientific knowledge that has specific commercial objectives with respect to new products, services, processes, or methods. Available online at http://www.caspar.nsf.gov/nsf/srs/IndRD/glossary.htm. Last accessed on November 16, 2004.

Inquiry aimed at gaining the knowledge or understanding to meet a specific, recognized need of a practical nature, especially needs to achieve specific commercial objectives with respect to products, processes, or services. Available online at http://energytrends.pnl.gov/glosn_z.htm. Last accessed on November 16, 2004.

Any research which is used to answer a specific question, determine why something failed or succeeded, solve a specific, pragmatic problem, or to gain better understanding. Available online at https://www.quirks.com/resources/glossary.asp. Last accessed on November 16, 2004.

Focused, systematic study and investigation undertaken to discover the applications and uses of theories, knowledge, and principles in actual work or in solving problems. See Research. Available online at http://www.siu.edu/orda/general/glossary.html. Last accessed on November 16, 2004.

Systematic study to gain knowledge or understanding necessary to determine the means by which a recognized and specific need may be met. [OMB Circular A-11, June 1996.] See also Conduct of Research and Development. Available online at https://radius.rand.org/radius/demo/glossary.html. Last accessed on November 16, 2004.

The investigation of some phenomena to discover whether its properties are appropriate to a particular need or want. In contrast, basic research investigates phenomena without reference to particular human needs and wants. Available online at http://www.unmc.edu/ethics/words.html. Last accessed on November 16, 2004.

Is designed to solve practical problems of the modern world, rather than to acquire knowledge for knowledge's sake. Available online at http://ventureline.com/glossary_A.asp. Last accessed on November 16, 2004.

Research designed for the purpose of producing results that may be applied to real world situations. Topic areas: Accountability and Evaluation. Available online at http://www.nonprofitbasics.org/SearchEntireSite.aspx?Source=2&SiteSearchText=research&PW=No&PreviousWord=research&C0=178&C4=1&C3=4&C5=36&C6=18&C1=92&C2=1. Last accessed on November 16, 2004.

Research that studies the relationship or applicability for theories or principles of a particular field to a particular problem. Available online at http://researchoffice.astate.edu/glossary_of_proposal_terms.htm. Last accessed on November 16, 2004.

Research done with the intent of applying results to a specific problem. Evaluation is a form of applied research. This can be conducted as part of an action research approach. Available online at http://www.sachru.sa.gov.au/pew/glossary.htm. Last accessed on November 16, 2004.

Research aimed at improving the quality of life and solving practical problems. Available online at http://highered.mcgraw-hill.com/sites/0072358327/student_view0/chapter1/glossary.html. Last accessed on November 16, 2004.

APPENDIX D

The aim is to address an immediate problem. The purpose is to try ideas in the context of educational (classroom) settings. Available online at http://www.ied.edu.hk/csnsie/ar/chap1/1_glossary.htm. Last accessed on November 16, 2004.

Applied research is that effort that (1) normally follows basic research, but may not be severable from the related basic research, (2) attempts to determine and exploit the potential of scientific discoveries or improvements in technology, materials, processes, methods, devices, or techniques, and (3) attempts to advance the state of the art. Applied research does not include efforts whose principal aim is design, development, or test of specific items or services to be considered for sale; these efforts are within the definition of the term development. Available online at http://www-agecon.ag.ohio-state.edu/class/AEDE601/glossary/glossa.htm. Last accessed on November 16, 2004.

Conducted to solve particular problems or answer specific questions. Available online at http://www.nelson.com/nelson/hmcanada/ob/glossary.html. Last accessed on November 16, 2004.

In applied research the objective of the sponsoring agency is to gain knowledge or understanding necessary to determine how a recognized need may be met. Available online at http://www.nsf.gov/sbe/srs/sfsucni/method99/help/glossary.htm. Last accessed on November 16, 2004.

Utilizing pure research to develop real-world products. Available online at http://www.investorwords.com/236/applied_research.html. Last accessed on November 16, 2004.

As opposed to basic research, applied research is the type of research which is conducted to solve practical problems, find cures to illnesses, develop therapies with the purpose of helping people, and other similar types of practical problem-solving research. Available online at http://www.alleydog.com/glossary/definition.cfm?term=Applied%20Research. Last accessed on November 16, 2004.

FUNDAMENTAL RESEARCH

National Security Decision Directives: "Fundamental research" means basic and applied research in science and engineering, the results of which ordinarily are published and shared broadly within the scientific community, as distinguished from proprietary research and from industrial development, design, production, and product utilization, the results of which ordinarily are restricted for proprietary or national security reasons. Available online at http://www.fas.org/irp/offdocs/nsdd/nsdd-189.htm. Last accessed on November 16, 2004.

Fundamental research is basic and applied research in science and engineering where the resulting information is ordinarily published and shared broadly within the scientific community. It is distinguished from proprietary research and from industrial development, design, production, and product utilizations, the results of which ordinarily are restricted for proprietary and/or specific national security reasons. Normally, the results of "fundamental research." are published in scientific literature, thus making it publicly available. Research which is intended for publication, whether it is ever accepted by scientific journals or not, is considered to be "fundamental research." A large segment of academic research is considered "fundamental research". Because any information, technological or otherwise, that is publicly available is not subject to the Export Administration Regulations (EAR) (except for encryption object code and source code in electronic form or media) and thus does not require a license, 'fundamental research' is not subject to the EAR and does not require a license. Available online at http://www.umbi.umd.edu/rcc/fundamentalresearch.pdf. Last accessed on November 16, 2004.

National Security Decision Directive 189: Fundamental Research defined: basic and applied research in science and engineering, the results of which are available to interested scientific community. National Policy: No restriction may be placed upon the conduct or reporting of federally funded Fundamental Research that has not received national security classification. This is reflected in ITAR at 22 CFRR 120.11(8). Executive Order 12356 (1985). Available online at http://www.epic.org/open_gov/eo_12356.html. Last accessed on November 16, 2004.

Federal Acquisition Regulation 27.404 (g) (2): In contracts for basic or applied research universities or colleges, no restrictions may be placed upon the conduct of or reporting on the results of unclassified basic or applied research, except as provided in applicable U.S. Statutes. Available online at http://supply.lanl.gov/Property/ecco/History/2004/presentations2004/default.shtml. Last accessed on November 16, 2004.

DoD Supplement to the FAR: It is DOD policy . . . to allow the publication and public presentation of unclassified contracted fundamental research results. The mechanism for control of information generated by DOT funded contracted fundamental research . . . is security classification. Available online at http://supply.lanl.gov/Property/ecco/History/2004/presentations2004/default.shtml. Last accessed on November 16, 2004.

OTHER

Defense of Basic Research by Joseph Henry: In 1852, Henry defended basic research. It was "profitable," he said, when that word was defined properly. "The true, the beautiful, as well as the immediately practical, are all entitled to a share of attention. All knowledge is profitable; profitable in its ennobling effect on the character, in the pleasure it imparts in its acquisition, as well as in the power it gives over the operations of mind and matter. All knowledge is useful; every part of this complex system of nature is connected with every other. Nothing is isolated. The discovery of to-day, which appears unconnected with any useful process, may, in the course of a few years, become the fruitful source of a thousand inventions." Available online at http://www.si.edu/archives/ihd/jhp/joseph04.htm. Last accessed on November 16, 2004.

Appendix E

Universities That Received Department of Defense 6.1 and 6.2 Funding in Fiscal Year 2002

Tables E-1 through E-3 show the top 50 university recipients of DOD 6.1, 6.2, and 6.1 + 6.2 funding in fiscal year 2002. The tables were constructed on the basis of data provided in personal communication from Mark Herbst, a member of the staff of the Office of the Director of Basic Research, within the Office of the Secretary of Defense, to James Garcia of the National Research Council. Institution names in **boldface** type and shaded background are those of universities at which committee members conducted site visits and interviews and/or held discussions with university research leaders who were not visited in person.

TABLE E-1 Top 50 University Recipients of DOD 6.1 Funding in Fiscal Year 2002

	Institution Name	State	Total 6.1 ($000)
1	**Massachusetts Institute of Technology**	MA	43,802
2	**Pennsylvania State University**, All Campuses	PA	35,357
3	**University of California, Los Angeles**	CA	31,784
4	**University of Washington**	WA	29,884
5	Stanford University	CA	25,811
6	**University of Southern California**	CA	25,758
7	Duke University	NC	25,607
8	**University of Michigan**, All Campuses	MI	24,245
9	University of California, San Diego	CA	24,117
10	**University of California, Santa Barbara**	CA	19,964
11	University of Illinois at Urbana-Champaign	IL	19,925
12	California Institute of Technology	CA	19,478
13	Georgia Institute of Technology, All Campuses	GA	19,098
14	Johns Hopkins University	MD	18,888
15	**University of Texas at Austin**	TX	17,044
16	Carnegie Mellon University	PA	16,527
17	**Cornell University**, All Campuses	NY	15,209
18	Princeton University	NJ	13,452
19	State University of New York, System Office	NY	12,598
20	Woods Hole Oceanographic Institution	MA	12,493
21	University of Pennsylvania	PA	12,282
22	Virginia Polytechnic Institute & State University	VA	11,476
23	**University of New Mexico**, All Campuses	NM	11,032
24	**University of Arizona**	AZ	10,977
25	University of Wisconsin - Madison	WI	10,389
26	University of Maryland System Administration	MD	10,334
27	University of Colorado, All Campuses	CO	10,303
28	Harvard University	MA	10,081
29	**Northwestern University**	IL	9,692
30	Purdue University, All Campuses	IN	9,479
31	**North Carolina State University at Raleigh**	NC	9,311
32	**Ohio State University**, All Campuses	OH	9,190
33	Arizona State University Main	AZ	9,051
34	University of Minnesota, All Campuses	MN	8,667
35	University of Virginia, All Campuses	VA	8,335
36	Rutgers, State University of New Jersey, All Campuses	NJ	8,262
37	University of Miami	FL	8,105
38	University of California, Davis	CA	7,776
39	University of California, Irvine	CA	7,672
40	Brown University	RI	7,491
41	Baylor College of Medicine	TX	7,040
42	Rice University	TX	6,918
43	San Diego State University	CA	6,893
44	University of North Carolina at Chapel Hill	NC	6,576
45	University of Pittsburgh, All Campuses	PA	6,305
46	New Mexico Institute of Mining and Technology	NM	6,129
47	Boston University	MA	6,005
48	University of South Carolina, All Campuses	SC	5,967
49	University of Florida	FL	5,947
50	University of Texas Southwestern Medical Center, Dallas	TX	5,862

SOURCE: Based on data provided in personal communication from Mark Herbst, Office of the Director of Basic Research, Office of the Secretary of Defense, Washington, D.C., to James Garcia, National Research Council, June 2004.

APPENDIX E

TABLE E-2 Top 50 University Recipients of DOD 6.2 Funding in Fiscal Year 2002

	Institution Name	State	Total 6.2 ($000)
1	**University of Southern California**	CA	32,586
2	University of Hawaii System Office	HI	20,131
3	**Pennsylvania State University**, All Campuses	PA	16,757
4	University of South Florida	FL	15,729
5	University of Dayton	OH	14,528
6	Carnegie Mellon University	PA	12,883
7	Georgia Institute of Technology, All Campuses	GA	10,429
8	Johns Hopkins University	MD	9,687
9	Auburn University, All Campuses	AL	7,572
10	University of Central Florida	FL	7,369
11	**Massachusetts Institute of Technology**	MA	7,206
12	**University of Texas at Austin**	TX	6,794
13	Vanderbilt University	TN	6,051
14	Florida State University	FL	6,000
15	**University of Washington**	WA	5,727
16	Stanford University	CA	5,274
17	University of Pennsylvania	PA	4,727
18	**Northwestern University**	IL	4,677
19	Florida Atlantic University	FL	3,640
20	University of Maryland System Administration	MD	3,519
21	**University of California, Los Angeles**	CA	3,441
22	University of California, Irvine	CA	3,426
23	University of Mississippi, All Campuses	MS	3,391
24	Boston University	MA	3,287
25	University of Illinois at Urbana-Champaign	IL	3,269
26	Washington State University	WA	3,221
27	University of Delaware	DE	3,051
28	**Cornell University**, All Campuses	NY	2,839
29	**Ohio State University**, All Campuses	OH	2,821
30	Rutgers, State University of New Jersey, All Campuses	NJ	2,709
31	New York University	NY	2,686
32	Utah State University	UT	2,519
33	University of California, San Diego	CA	2,513
34	Washington University	MO	2,388
35	University of Rhode Island	RI	2,322
36	University of Massachusetts at Amherst	MA	2,282
37	**University of California, Santa Barbara**	CA	2,202
38	Colorado State University	CO	2,090
39	Purdue University, All Campuses	IN	2,015
40	University of California, Riverside	CA	1,996
41	University of North Carolina at Charlotte	NC	1,980
42	Georgetown University	DC	1,922
43	Princeton University	NJ	1,909
44	University of Texas at Dallas	TX	1,879
45	Yale University	CT	1,854
46	Columbia University	NY	1,821
47	**University of Arizona**	AZ	1,801
48	Rensselaer Polytechnic Institute	NY	1,789
49	University of Colorado, All Campuses	CO	1,777
50	Drexel University	PA	1,694

SOURCE: Based on data provided in personal communication from Mark Herbst, Office of the Director of Basic Research, Office of the Secretary of Defense, Washington, D.C., to James Garcia, National Research Council, June 2004.

TABLE E-3 Top 50 University Recipients of DOD 6.1 + 6.2 Funding in Fiscal Year 2002

	Institution Name	State	Total 6.1 + 6.2 ($000)
1	**University of Southern California**	CA	58,343
2	**Pennsylvania State University, All Campuses**	PA	52,114
3	**Massachusetts Institute of Technology**	MA	51,007
4	**University of Washington**	WA	35,611
5	**University of California, Los Angeles**	CA	35,225
6	Stanford University	CA	31,085
7	Georgia Institute of Technology, All Campuses	GA	29,527
8	Carnegie Mellon University	PA	29,410
9	Johns Hopkins University	MD	28,575
10	University of California, San Diego	CA	26,629
11	**University of Michigan, All Campuses**	MI	25,935
12	Duke University	NC	25,774
13	**University of Texas at Austin**	TX	23,838
14	University of Illinois at Urbana-Champaign	IL	23,194
15	University of Hawaii System Office	HI	22,269
16	**University of California, Santa Barbara**	CA	22,166
17	University of South Florida	FL	20,409
18	California Institute of Technology	CA	20,363
19	**Cornell University, All Campuses**	NY	18,048
20	University of Dayton	OH	17,762
21	University of Pennsylvania	PA	17,009
22	Princeton University	NJ	15,361
23	**Northwestern University**	IL	14,369
24	State University of New York System Office	NY	13,969
25	University of Maryland System Administration	MD	13,853
26	Woods Hole Oceanographic Institution	MA	13,549
27	University of Central Florida	FL	13,224
28	**University of Arizona**	AZ	12,778
29	Virginia Polytechnic Institute and State University	VA	12,221
30	University of Colorado, All Campuses	CO	12,080
31	**Ohio State University, All Campuses**	OH	12,011
32	University of Wisconsin-Madison	WI	11,990
33	**University of New Mexico, All Campuses**	NM	11,869
34	Purdue University, All Campuses	IN	11,494
35	Harvard University	MA	11,346
36	University of California, Irvine	CA	11,098
37	Rutgers, State University of New Jersey, All Campuses	NJ	10,971
38	Vanderbilt University	TN	10,643
39	Auburn University, All Campuses	AL	10,484
40	**North Carolina State University at Raleigh**	NC	10,168
41	Arizona State University Main	AZ	9,893
42	University of Miami	FL	9,682
43	University of Minnesota, All Campuses	MN	9,333
44	Boston University	MA	9,292
45	University of Virginia, All Campuses	VA	9,220
46	Brown University	RI	9,147
47	Florida State University	FL	9,135
48	University of Delaware	DE	8,313
49	University of California, Davis	CA	7,876
50	Yale University	CT	7,700

SOURCE: Based on data provided in personal communication from Mark Herbst, Office of the Director of Basic Research, Office of the Secretary of Defense, Washington, D.C., to James Garcia, National Research Council, June 2004.